Animated Algorithms

**A Self-Teaching Course in Data Structures
and Fundamental Algorithms**

Michael P. Barnett

Simon J. Barnett

McGraw-Hill Book Company

New York St. Louis San Francisco Auckland Bogotá
Hamburg Johannesburg London Madrid
Mexico Montreal New Delhi Panama
Paris São Paulo Singapore
Sydney Tokyo Toronto

For Sally and Barbara

Library of Congress Cataloging-in-Publication Data

Barnett, Michael P.
  Animated algorithms.

  Includes index.
  1. Data structures (Computer science)   2. Algorithms.
I. Barnett, Simon J.   II. Title.
QA76.9.D35B365   1986   005.7′3   85-29749
ISBN 0-07-003792-2
ISBN 0-07-003797-3 (pbk.)

1234567890   DOC/DOC   8932109876

ISBN 0-07-003792-2   HC

ISBN 0-07-003797-3   SC

The editors for this book were Stephen G. Guty
and Galen H. Fleck, the designer was Naomi Auerbach,
and the production supervisor was Sally Fliess.
It was set in Century Schoolbook by Achorn Graphics.
Printed and bound by R. R. Donnelley & Sons Company.

Contents

Preface

Data structures and algorithms are at the heart of an ever-increasing variety of computer applications in business and education, the natural and social sciences, and the arts and humanities, as well as being staple elements of the computer science curriculum. People who use computers in inventory and traffic control, metabolic studies, instructional and bibliographic publishing, genealogy reconstruction, graphics and animation, language processing, and artificial intelligence all make extensive use of data structures and algorithms.

This book provides simple explanations and practical information for the working professional who needs to use data structures and algorithms and for the data processing manager who wants a better idea of what computer science provides to increase productivity. Over twenty ready-to-run BASIC programs let you watch the step-by-step progress of algorithms that sort and search lists of data, that zoom and pan through menus and data bases to help you browse and retrieve information, that rearrange and expand text and symbolic material systematically, and that explore paths around networks of facts and ideas. The language of data structures and algorithms is explained in simple terms, and it is related to a diversity of practical applications.

The animated displays provided by the programs in this book make the operation of the algorithms very clear. They also help the reader to develop insights into processes that are performed dynamically by the unidirectional and bidirectional streaming of information. This bears on the use and development of network, array, and parallel-processing systems and of dataflow concepts in general.

Many of the display techniques that we use can be adapted to other instructional situations that are of interest to readers concerned with the design and development of courseware.

Practical experience with data structures and algorithms in several areas of numeric and nonnumeric computation has influenced the selection of the examples that are explained in detail and the suggested activities that are outlined for the reader to pursue. Further experience with teaching data structure courses and supervising projects in which special structures are used has brought home the need for numerous and extensive worked examples. The programs in this book let readers experiment with as many examples as they wish while building confidence and reinforcing their basic understanding.

Although this book has been written as a self-contained introduction to the subject, it can be read with advantage in conjunction with a formal text and with some of the more advanced and specialized literature in the field. In my efforts over the years to teach courses on data structures and to widen my horizons, I have found much of this material invaluable. Some of it is mentioned in the Bibliography. The students and colleagues who have helped shape my ideas are far too numerous to mention individually, but I thank them all the same.

The programs in this book run on any of the IBM PC family of computers equipped with 128K and a graphic card. They also run on many IBM PC compatibles. To purchase the programs on disk, write to TYCON, P.O. Box 3718, Princeton, NJ 08543.

Michael P. Barnett

Introduction

Most of this book deals with data structures and algorithms that support three major kinds of activity:

1. Processes to sort, search, alter, and combine arrays and files of data and to use this data sequentially. These processes deal with lists.
2. Processes that, for example, branch to subroutines and find the correct point of return afterward. These processes have the common property of digressing from one stream of data or instructions to another, returning to the appropriate point when each digression is complete, and using *stacks* to keep track of the backtracking information.
3. Processes that follow paths through networks and other interconnected bodies of data called *trees* and *graphs*.

These three kinds of processes overlap and support each other. In fact, it is probably more important to stress the coherence of the field of data structures and algorithms than to try to divide it into separate compartments. The development of topics in the successive chapters really unfolds an integrated methodology with several themes recurring as the tale progresses.

Previewing the structure in a little more detail, Chapter 1 deals largely with sorting. It begins with a likely point of contact for many readers, the bubble sort. This is discussed in conjunction with some other elementary sorting algorithms. The chapter goes on to pointer sorting and to the quicksort algorithm that is better than the elementary algorithms for many purposes. The chapter ends with some comments on how different methods of sorting can be assessed and some practical comparisons and suggestions.

Chapter 2 deals with binary searching, the use of directories that are akin to multilevel tables of contents, and processes to merge and combine files of data in related ways. Individual topics in this chapter apply to the mainstream of conventional data processing, to information storage and retrieval, and to operations on sets of data (where we use the word "set" with its mathematical meaning). The chapter ends with a discussion of techniques to access material in lengthy files. The example that uses directories shows how to work with several levels of window. The animated merge example conveys the visual impression of data flowing dynamically.

Chapter 3 explains the linked list technique that avoids the continual rearrangement of data in processes that change the content of lists frequently. The first program shows how a linked list keeps track of order. The second program shows how space can be reused after items have been deleted. The third program combines these processes, and the chapter ends with an analysis of several benefits of linked lists and related techniques for the management of storage.

Chapter 4 introduces the varied uses of stacks. It begins with some simple examples to reverse streams of data and to handle digressions such as those that occur when a program invokes a subroutine. Then some matters that are basic to the ability of computers to handle computations conveniently are discussed; they include the evaluation of postfix expressions, the conversion of algebraic expressions to postfix form, and the provision for parentheses in the input. A simple queue maintenance program is described; it demonstrates some of the basic principles of simulation. The chapter ends with a brief discussion of the use of stacks to handle recursion.

Chapter 5 describes some of the ways in which trees and graphs are stored and processed. We begin with the *structure table* representation of trees that provides a very natural approach to many kinds of applications and is easy to follow and to modify to fit details of the data. The processes to convert a general tree to a binary tree and to traverse a binary tree are explained. Graphs are introduced by a program to find the shortest path between any two points in a network. The chapter ends with a brief description of some of the special kinds of trees that have been studied extensively—to expedite searching processes, for example—and with some comments on the important areas of graph processing.

Whatever your field of interest, the methods of Chapters 1 to 5 match a succession of applications that you may wish to pursue. Business data processing includes sorting, searching and retrieving sales, personnel, supplier, manufacturing, and advertising information. Real-time transaction processing uses linked lists. Econometric techniques to minimize shipping costs and to site plants and warehouses use stacks, graphs, and trees.

Scientists and engineers sort and merge experimental data and search numerical and bibliographic data bases when designing products and planning and interpreting laboratory work. Chemical formulas, anatomical and geological structures, metabolic pathways, and engineering structures can be processed by techniques for handling graphs and trees.

In the arts, humanities, and social sciences, too, graph-processing methods are applied to depictions of archeological sites, architectural plans, choreographic designs, museum displays, patterns of causality and literary influence, kinship structures, and linguistic evolution. Lawyers use graphs of corporate relations and, in cross-examination, trees of questions.

For experts in these various fields, the literature of data structures and algorithms provides an embarrassment of riches. To the lawyer, architect, chemist, scenic designer, and business executive with a personal computer and a working knowledge of BASIC who wants to represent a particular body of information in a manner that can be understood and processed easily, the standard discussions of O(n log n) search performance, of trees with names that are as cryptic to the uninitiated as Latin tags in an arboretum, and the allusions to topological problems and recursive function theories of great profundity—all these can be intellectually challenging but pragmatically discouraging.

The programs in this book illustrate numerous techniques that are of interest in the design of courseware in general and, in particular, that demonstrate step-by-step cycling through algorithmic processes. They include tactics to focus attention on items of current interest, to show changing paths and patterns of connection, and to make the prompts provide running commentaries about the processes that are being demonstrated. Clutter and discord are avoided in the displays, and the juxtaposition of "before" and "after" states of the data provides continuity.

Particular display requirements sometimes control the overall design of a demonstration. Thus, to follow the application of the quicksort algorithm to a particular list, it helps to be able to look back at the recent history of the comparisons and transpositions that were made. For this demonstration to be effective, it seems necessary, too, to be able to use rather more items in the list than are needed to convey the principle of the bubble sort. By using the text mode with 80 characters across the screen and displaying the list horizontally, 25 two-digit numbers can be shown in a single line with brackets to delimit the sublist under current attention. To demonstrate the principle of linked lists, however, the main needs seem to be (1) the use of a list of words that are at least three or four letters long and (2) the provision of enough space between the words to let link lines "snake around" from the pointer that is displayed on the right of one item to the slot number that is displayed at the left of another item. Twelve items seem quite enough for a viable demonstration of the linked list principle. The programs in Chapter 3, accordingly, display the lists vertically, and blank lines are left between the items for the link lines to be drawn when necessary.

Particular sections of many of the programs serve as prototypes of what we may call programming gambits or unit processes by analogy to the use of similar terms in board games and in chemical engineering. They include changing the color of the text in medium-resolution graphics, drawing symmetrical arrowheads on lines that point in arbitrary directions in high resolution, displaying a value between parentheses without intervening spaces, drawing rectilinear paths and outlines with graphics characters and cursor control codes, moving data around the display with GET and PUT statements, using tiling patterns and line textures to distinguish areas and paths of different status, and using character strings to store and manipulate lists of small integers. Prompts, responses, error trapping, diagnostics, and remedial and retry actions are handled in systematic ways. These and other similar matters are discussed in the text, mentioned in REMs in the program listings, and cited in the indexes.

Each section in this book deals with a separate program and begins with a description of what the user sees when the program is run, including an account of the prompts and the behavior triggered by different responses. The variables used in the program are listed and explained, and the key sections of the program that merit comment are discussed. Each section ends with a set of suggested activities based on the program and a program printout containing extensive REMs.

Some Simple Sorting Processes

People have been putting things in order since the dawn of history and probably for quite a while before then, too. The need to sort census data led to the invention of punched-card machines in the nineteenth century. These machines paved the way for the modern data processing industry, and sorting has played a key role in computer applications of every kind. Whatever use you make of computers, sorting is important to you for several reasons. Data can be found for inspection and alteration and checked for completeness more easily in a sorted file than in a random file. Systematic processing of all the records in a file often requires that the records be in a particular order. Also, comparing or combining the information in a pair of files requires that the two files have been sorted in the same way.

The item that determines the order of the records in a sorted file is called the *key*. Often, it is a number that has to increase from record to record. Alternatively, the key may be a character string such as a name, and then the file is sorted to put the character strings for all the records in alphabetic order. Occasionally, each record in a file consists of a single item; then the entire record serves as the key. At other times, the key is a property of the record, such as its length, that the computer can find even though it is not included in an actual field.

Most sorting operations process a file completely. Sometimes, however, the process is limited to finding a few items that would be at the start or the end of the list if it were sorted in full. Often, the purpose of the sort is to arrange information for later use inside the computer. In other applications, computers sort data that goes into a publication such as a directory or a part of a publication such as an index. Sometimes, data is consolidated, for example by printing 1–4 instead of 1, 2, 3, 4. At times, people need the sorted information to arrange some objects outside the machine; for example, when the computer is used to sort railroad cars to organize a freight train.

If you write or adapt programs for applications in which you are interested, sooner or later you will find the need to sort. At that time, the easiest tack is to choose an algorithm that you already know, that is easy to follow, and is short to code. If the amount of data is small, it will not matter if the algorithm is inefficient. When the data is more extensive, however, efficiency can be of critical importance. Sections 1.1 to 1.4 describe four programs that demonstrate several sorting principles, and Section 1.5 discusses some further ramifications of the subject.

1.1   Three Simple Schemes for Sorting

The program SIMPSORT demonstrates the three elementary sorting algorithms that are often called *bubble sort, insertion sort,* and *selection sort.* The bubble sort is probably the best-known computer algorithm, and many books on introductory programming and computer literacy include it. The insertion sort corresponds to the method that many people use to arrange a bridge or gin rummy hand. The selection sort also is very simple. The insertion sort is the best of these for many purposes, but the other two sorts have their merits and deserve attention.

The program SIMPSORT, which is displayed at the end of this section, begins by prompting

```
              Simple sort demonstration
    How many numbers do you want to sort (limit is 10):
```

You respond by typing a number between 1 and 10 and pressing the Enter key. The message on the bottom line of the screen changes to

```
    Press R for a random list, T to type the data, P for previous list:
```

If you press R, the program displays a list of random numbers in the center of the screen. For example, if you type 3 for how many you want sorted, the program displays the column of three numbers shown in (*a*). The program always gives the random number generator the same seed number. Consequently, the sequence of numbers that it provides is always the same.

716	716 swap	684	684	684	483

684	684	716	716 swap	483	684

483	483	483	483	716	716
(a)	(b)	(c)	(d)	(e)	(f)

After displaying the column of numbers, the program prompts:

```
Press B for bubble, I for insertion, S for selection sort:
```

We will describe the behavior for the bubble sort first. After you press B, the message

```
Press the space bar to proceed from step to step
```

appears at the bottom of the screen. The top two numbers are highlighted to show that the program is about to check if they are in the correct order. When you press the space bar, the word "swap" appears alongside the first number because that number is larger than the next number (b). The word also is highlighted, that is, displayed in black on white. This style of highlighting on the screen is shown by boldface type.

Now press the space bar again. The word "swap" disappears. The first two numbers are interchanged, and they remain highlighted (c). Press the space bar again. The number that is now at the top reverts to white on black. The numbers on the second and third lines are highlighted to show that they will be compared with each other. Press the space bar. The word "swap" appears on the second line because the second number is larger than the third (d). Press the space bar twice again. The word "swap" disappears and the numbers in the second and third lines are interchanged. Then the program displays a line of hyphens above the bottom number. This shows that the bottom number is in the final position (e). Whatever numbers you started with, the actions up to this point ensure that the largest number is now at the end.

Press the space bar. The process starts again from the top. The numbers in the first and second lines are highlighted. The word "swap" appears, because 684 is greater than 483. Accordingly these two numbers are interchanged. A line of hyphens appears above the second number to show that it also is in its final place (f).

At this point, the second number would be in its final place whatever numbers you started with and whether there had been any interchange of numbers or not. Because you started with just three numbers, the number in the first line must be the smallest, and therefore the sort is complete. The line of hyphens disappears, and the prompt for further action appears beneath the sorted list:

```
Press A for another example, Q to quit:
```

If you press A, the screen clears and the original message appears at the bottom of the screen; it asks how many numbers you want to sort. If you enter the value 4 and press R in response to the next prompt, the program displays another list that is produced by the random number generator. This second list consists of the numbers 1000, 647, 133, 370, displayed vertically.

You are then prompted to specify the sorting process. If you type B for bubble again and press the space bar repeatedly, you will see the number 1000 swapped with every other number in turn and made to move down repeatedly until it reaches the bottom of the list. If it had been at an intermediate point to begin with, then it would have remained in place when it was used as the second number in a swap or not-to-swap comparison (because, being largest, whatever was above it would be smaller). Thereafter, it would have been swapped with each further number in turn until it reached the bottom. If the largest number had been at the bottom to begin with, then, whatever happened before it was reached as the second item in a comparison, the number above it at that time would have been smaller and the largest number would have stayed in place.

To go on with the example at hand, the four numbers are now in the order 647, 133, 370, 1000, reading down the screen. Pressing the space bar a few more times swaps the number 647 with the numbers 133 and 370, which puts the list in the order 133, 370, 647, 1000. A line of hyphens is displayed under the number 370 to show that the portion of the list beneath this number is in order. Thus, at the end of the second cycle the two largest numbers are in the two bottom positions of the display. As an intelligent and observant human being, you can see that the entire list is now ordered. The very simple program that we are using, however, makes one more sweep to move the third largest number to the third position from the end or to leave it in place if it is already there. Then the line of hyphens disappears and you are prompted for further action in the manner described.

You can proceed to display and sort as many sets of random numbers as you wish. If, however, you type T to input your own data, the program prompts for 10 integers between −9999 and 99999. The restriction is only for reasons of simplicity; waiving it would just make the formatting action in the program a little more complicated. You respond by typing 10 numbers. The cursor automatically moves to the center of successive lines of the screen for the successive items, and it adjusts their horizontal positions to right-align them.

After you have entered the list of numbers, the program prompts for the first letter of the sorting algorithm. You press the space bar to step through the process. When the sort is complete, the prompt to type A or Q reappears. You can run more examples by using your own data, or you can take things easy and let the computer provide random numbers for you. You can also repeat an example any time you wish by typing P in response to the prompt for the source of data. When you complete a sort and want to quit, type Q and the Ok message appears in the top-left corner of the screen.

The bubble sort algorithm got its name from the fact that the smallest numbers rise to the top like bubbles in a bath. Correspondingly, the larger numbers sink, but to call a process the sinking algorithm would not fit in with popular expectations of positive thought. There are several ways to streamline the version that we have just described. Some of them are included in the Suggested Activities at the end of the section.

Now suppose that you type I in response to the appropriate prompt to make the program apply the insertion sort algorithm to the list 716, 684, 483. Initially, this has the appearance shown in (*g*). Pressing the space bar repeatedly produces the following actions. The 716 and 684 are highlighted to show that they are being compared. A side comment appears to tell you so. A line of six hyphens appears below the 684 for later reference (*h*).

```
716                      716

684                      684  compare 684 and 716
                         ------
483                      483
(g)                      (h)
```

Then the program prepares to move the 684 to an earlier position in the list, because it is smaller than the 716. The display changes in turn to:

```
716                   716  move 716 down        insert 684 here

   move 684 back          move 684 back         716
------                    ------                ------
483                   483                        483
(i)                       (j)                    (k)
```

```
684                   684

716                   716
------
483                   483  compare 483 and 716
                      ------

(l)                   (m)
```

Thus, the 684 is slid out (i) and the 716 is slid down (j) and (k). The 684 is slid into the place that the 716 occupied (k) and (l), because there are no slots any further back. The 716 and the 483 are highlighted (m) to show that they will be compared next. The display changes in turn to

```
684                   684                        684

716                   716 move 716 down              compare 483 and 68
                                                 716
      move 483 back         move 483 back
------                ------                     ------

(n)                   (o)                        (p)
```

Thus, because the 483 is smaller than its predecessor, it is slid out (n). The 716 is moved down again (o) and (p), and the highlighting is switched to the 684 to show it is being compared with the 483 (p). Because 483 is less than 684, the 684 is moved down (q) and (r). Then, because there are no items to check further back, the 483 is slid into the position that the 684 occupied (r) and (s).

```
684 move 684 down                insert 483 here      483

     compare 483 and 684         684                  684

716                              716                  716
------                           ------               ------

(q)                              (r)                  (s)
```

The program has now taken care of the entry that ended the original list. Consequently, the sort is complete.

Now suppose that you use the insertion sort on the list 1000, 647, 133, 370. In the first cycle, the entries in the first and second slots are compared. The 647 is slid out, because it is less than 1000. The 1000 is slid down, and the 647 is slid into the slot that the 1000 occupied. This puts the elements in the order 647, 1000, 133, 370. In the next cycle, the 1000 and 133 are compared. The 133 is slid out, the 1000 and 647 are slid down, and the 133 is slid back in. The list at this point is in the order 133, 647, 1000, 370. In the next cycle, the 1000 and the 370 are compared. The 370 is slid out. The 1000 and 647 are moved down. Then, because 370 is more than 133, the 370 is slid into the position that the 647 just occupied. Because the 370 has been processed, the sort is complete. If you have ever played cards, you can see that this is the process you use when you look at the cards in your hand from left to right, remove the first card that is out of order, and look back for the correct place to put it. Whenever the insertion sort option of SIMPSORT finds two items already in order at the start of a cycle, it displays "do not move," and goes on to the next cycle.

Now suppose that you type S in response to the prompt for the choice of algorithm to apply the selection sort to the list of three items that was used to explain the bubble sort and the insertion sort. The display, after successive key presses, is as follows:

```
716 largest in current pass so far     716 largest in current pass so far

684                                     684 smaller

483                                     483

(t)                                     (u)


716 largest in current pass so far

684

483 smaller

(v)
```

The program has reached the final slot, with 716 found to be the largest item in the current cycle. Accordingly, the 716 is swapped with the element in the final slot. The display changes in turn to:

```
716 largest in current pass so far      483              483

684                                      684              684
                                         ------           ------
483 swap                                 716              716
(w)                                      (x)              (y)
```

Thus, the first cycle ensures that the largest item is at the end, as in the bubble sort. The program now searches for the largest item that is still above the hyphens. The display changes in turn to:

```
483 largest in current pass so far      483 largest in current pass so far

684                                      684 larger
------                                   ------
716                                      716
(z)                                      (aa)
```

483		483		483

684	largest in current pass so far	684	in place already	684
------		------		
716		716		716
(*bb*)		(*cc*)		(*dd*)

The program then recognizes that only one item is left above the hyphens and, therefore, that the sort is complete. The hyphens disappear, and the program prompts for further action. A number is also made the "largest in current pass so far" when it is found to be equal to the number that previously had that status.

When you respond to a prompt from SIMPSORT or from any of the other programs in this book that require a single letter, such as B, I, or S, to specify the kind of sort, you may type either the capital or the lowercase letter. For most of the prompts, too, the first option that a prompt offers is also obtained by pressing any other key. Thus pressing any key in response to the request to type B, I, or S has the same effect as pressing B. Also, whenever we refer to pressing the space bar to make the program go on to the next step, pressing any other key will do just as well. If you respond to a prompt for data by typing a value that the program does not accept, a displayed diagnostic message will tell you to press the space bar and type a value that is allowed.

The relative merits of the three algorithms are discussed in Section 1.5, at the end of the chapter.

Program Variables

X(k)	The item that currently occupies the kth position in the list that is being sorted
P(k)	The kth item in the array that preserves the starting values of the current contents of the array X
M	The number of integers to be sorted
I	The index of the loops that construct the array X and that copy it
Q	The index of the item that is about to be displayed or for which a side comment is to be displayed or deleted
D	The index of a delay loop that slows down the display of successive items
J	The index of the outer loop in each of the sorting processes
K	The index of the inner loop
XC	The entry that is being moved into place, relative to the previous entries, in the current cycle of the insertion sort algorithm
XG	The largest entry found so far in the current outer cycle of the selection sort algorithm

KG   The index of this entry

R$ The one-character response to the prompts for an option that is specified by a single letter

Z$ The one-character response to the request for a key press to proceed from step to step (this convention is used in all the programs)

Program Notes

The section that generates random numbers between 1 and 1000 uses the statement

```
430  X(I)=1+INT(1000*RND)
```

This works as follows. RND generates random numbers. The smallest value that the system lets RND produce is 0; the largest value is slightly less than 1. Consequently, the smallest possible value of 1000*RND is 0, and the largest is slightly less than 1000. The INT function of 0 is 0; the INT function of 999.99999 is 999. The smallest value of the entire expression is $1+0=1$; the largest value is $1+999=1000$.

The statement

```
410  :'RANDOMIZE TIMER
```

is treated as a remark because of the :' symbols. As a result, the random number generator always begins with the same initial value when this program is run, and the numbers used in the explanation at the start of this section always appear. To make the program generate different sets of random numbers whenever it is restarted, delete the :' to activate the RANDOMIZE TIMER statement.

The bubble sort algorithm is applied by a nested loop. The index J of the outer loop limits the cycling of the inner loop.

```
1030 FOR J=M to 2 STEP -1
1040   FOR K=1 TO J-1
. . .
1290   NEXT K
. . .
1530 NEXT J
```

Within this nest, the numbers are compared and, when necessary, interchanged by

```
1210 IF X(K)<=X(K+1) THEN 1250
1220   SWAP X(K),X(K+1)
```

```
100 REM * SIMPSORT demonstrates bubble, insertion and selection sorts *
109 :
110   ON ERROR GOTO 6000                                         'error trap
120   SCREEN 0: KEY OFF                                 'set up the screen
130   COLOR 7,0: WIDTH 80
140   DIM X(10),P(10)
199 :
200 REM * start a demonstration *
210   CLS                                             'clear the screen
220   LOCATE 23,27                                           'caption
230    PRINT "Simple sort demonstration";
240   LOCATE 25,14                                             'prompt
250   INPUT; "How many numbers do you want to sort (limit is 10): ",M
260    IF 1<M AND M<=10 THEN 320
270      LOCATE 25,14                                 'invalid response
280        PRINT SPC(65)
290      LOCATE 25,20
300        INPUT; "Please type a number between 2 and 10: ",M
310      GOTO 260
320   LOCATE 25,1                                             'prompt
330    PRINT "Press R for a random list, ";
340    PRINT "T to type the data, P for previous list:" TAB(79)
350   R$=INPUT$(1)                                    'store the response
360   CLS
370   ON INT((1+INSTR("RrTtPp",R$))/2) GOTO 400,500,800         'branch
399 :
400 REM * program generates random numbers *
410   :'RANDOMIZE TIMER                    'different sequence each time
420   FOR I=1 TO M
430    X(I)=1+INT(1000*RND)                    'number between 1 and 100
440     Q=I: GOSUB 5000                                    'display it
450     FOR D=1 TO 100: NEXT D                                 'pause
460   NEXT I
470    GOTO 900
499 :
500 REM * user provides the numbers *
510   LOCATE 24,16
520    PRINT "Please type" M "integers between -9999 and 99999";   'prompt
530    FOR I=1 TO M
540     LOCATE 2*I-1,37
550      INPUT "",X(I)                            'store the ith number
560       X(I)=INT(X(I))                    'truncate if not an integer
570       IF -9999<=X(I) AND X(I)<=99999! THEN 650   'check it is in range
580        LOCATE 2*I,6
590          PRINT "Please stay between -9999 and 99999. ";    'diagnostic
600           PRINT "Press space bar and try again."
610          Z$=INPUT$(1)                              'wait for key press
620          LOCATE 2*I-1,1
630           PRINT SPACE$(160);                    'erase the diagnostic
640         GOTO 540
650       LOCATE 2*I-1,37
660        PRINT SPC(40)             'clear the line on which it was typed
670       Q=I: GOSUB 5000                      'redisplay it right aligned
680     NEXT I
690    GOTO 900
799 :
800 REM * rerun the previous list *
805   M=MP                              'use length of preserved list
810   FOR I=1 TO M
820    X(I)=P(I)                    'copy an item from the preserved list
```

```
830    Q=I: GOSUB 5000                                              'display it
840    NEXT I
850    GOTO 900
899  :
900  REM * prepare to sort *
910    FOR I=1 TO M                                      'preserve the current list
920    P(I)=X(I)
930    NEXT I
935    MP=M'                                           preserve the length
940    LOCATE 24,1                                                   'prompt
950    PRINT "Type B for bubble, I for insertion, S for selection sort:" TAB(79)
960    R$=INPUT$(1)
970    ON INT(INSTR("BbIiSs",R$)/2) GOTO 1000,2000,3000             'branch
999  :
1000 REM * bubble sort *
1010   LOCATE 24,1
1020    PRINT TAB(15) "Press the space bar to proceed from step to step.";
1030   FOR J=M TO 2 STEP -1                                       'outer loop
1040    FOR K=1 TO J-1                                            'inner loop
1100 REM * highlight the entries to be compared *
1110     Q=K: GOSUB 5100                                  'highlight Kth item
1120     Q=K+1: GOSUB 5100                        'highlight the (K+1)th item
1130     Z$=INPUT$(1)                                  'wait for key press
1140     LOCATE 2*K-1,42
1150     IF X(K)<=X(K+1) THEN PRINT " do not";   'display do not swap or swap
1160      PRINT " swap";
1170     Z$=INPUT$(1)                                  'wait for key press
1200 REM * swap if necessary *
1210     IF X(K)<=X(K+1) THEN 1250                'jump if items in order
1220     SWAP X(K),X(K+1)                                'swap them otherwise
1230     Q=K: GOSUB 5100                                 'redisplay them
1240     Q=K+1: GOSUB 5100
1250     Q=K: GOSUB 5200                                 'erase side comment
1260     Z$=INPUT$(1)
1270     Q=K: GOSUB 5000                 'de-highlight the last two items
1280     Q=K+1: GOSUB 5000
1290    NEXT K
1500 REM * end an outer loop *
1510    Q=J: GOSUB 5300                         'mark position with hyphens
1520    Z$=INPUT$(1)
1530   NEXT J
1540   GOTO 4000
1999 :
2000 REM * insertion sort *
2010   LOCATE 24,1
2020    PRINT "Press the space bar to proceed from step to step." TAB(79)
2030   FOR J=2 TO M                                              'outer cycle
2040 REM * highlight the entries to be compared at start of sweep *
2050     Q=J-1: GOSUB 5100                                     'highlight
2060     Q=J: GOSUB 5100
2070     XC=X(J)                                   'entry to be placed
2080     COLOR 7,0
2090     IF J=1 THEN 2120
2100      LOCATE 2*J-2,32                   'mark starting position with hyphens
2110       PRINT SPC(6)
2120      LOCATE 2*J,32
2130       PRINT "------"
2140     COLOR 0,7                               'display side comment
2150     LOCATE 2*J-1,42
2160      PRINT " compare" XC "and" X(J-1);
```

```
2170     Z$=INPUT$(1)
2180     GOSUB 5200                                    'erase side comment
2190     IF XC<X(J-1) THEN GOTO 2310                   'branch on comparison
2200 REM * the entries are in the correct order *
2210     COLOR 0,7                                     'display side comment
2220      LOCATE 2*J-1,42
2230       PRINT " do not move"
2240     Z$=INPUT$(1)
2250      Q=J: GOSUB 5200                              'erase side comment
2260      Q=J-1: GOSUB 5000                            'de-highlight the entries
2270      Q=J: GOSUB 5000
2280     GOTO 2710
2300 REM * an entry must be moved back *
2310     K=J                                           'prepare to search backwards
2320      COLOR 0,7                                    'display side comment
2330       LOCATE 2*J-1,32
2340        PRINT STRING$(10,219) " move" XC "back"
2400 REM * cycle backwards *
2410     WHILE XC<X(K-1) AND K>1                       'search for insertion point
2420      Q=K-1: GOSUB 5100                    'highlight the entry being compared
2430      Z$=INPUT$(1)
2440       LOCATE ,42                                  'display side comment
2450        PRINT "move" X(K-1) "down"
2460      Z$=INPUT$(1)
2470      X(K)=X(K-1)                                  'slide an entry down
2480      Q=K-1: GOSUB 5400                    'delete from prior position
2490      Q=K: GOSUB 5000                      'display in new position
2500      GOSUB 5200                                   'erase side comment
2510      K=K-1                                        'decrement the pointer
2520      IF K=1 THEN 2560
2530       COLOR 0,7                                   'display side comment
2540        LOCATE 2*K-1,42
2550         PRINT "compare" XC "and" X(K-1);
2560     WEND
2570     IF K>1 THEN Z$=INPUT$(1)
2600 REM * insert the entry that had to be moved back *
2610      COLOR 0,7                                    'display side comment
2620       LOCATE 2*K-1,42
2630        PRINT "insert" XC "here" STRING$(5,219)
2640      Z$=INPUT$(1)
2650     X(K)=XC                             'insert the entry that is being placed
2660      Q=K: GOSUB 5000                                    'redisplay it
2670      GOSUB 5200                                   'erase side comment
2680      Z$=INPUT$(1)
2690      Q=J: GOSUB 5200                              'erase side comment
2700      Q=J+.5: GOSUB 5400                           'erase marker hyphens
2710     NEXT J
2720     GOTO 4000
2999 :
3000 REM * selection sort *
3010      LOCATE 24,1
3020       PRINT TAB(15) "Press the space bar to proceed from step to step.";
3030     FOR J=1 TO M-1                                'outer loop
3040      KG=0                                 'initialize pointer and value
3050      XG=0                                         'of largest item so far
3100 REM * inner loop *
3110      FOR K=1 TO M-J+1                             'find largest item
3120       Q=K: GOSUB 5100                             'highlight an entry
3130       IF K=1 THEN 3350
3140        Z$=INPUT$(1)
```

```
3150      IF X(K)>=XG THEN 3310                              'jump if larger or equal
3160      LOCATE 2*K-1,42                                     'display side comment
3170       PRINT " smaller"
3180      Z$=INPUT$(1)
3190      IF K<M-J+1 THEN Q=K: GOSUB 5000: PRINT SPC(35)        'de-highlight
3200      GOTO 3410
3300 REM * current item is larger *
3310      LOCATE 2*K-1,42                                     'display side comment
3320      IF X(K)>XG THEN PRINT " larger" ELSE PRINT " equal"
3330      Z$=INPUT$(1)                                      'remove comment from
3340      IF KG<M-J+1 THEN Q=KG: GOSUB 5000: PRINT SPC(35)   'previous largest
3350      XG=X(K)                                            'update largest value
3360      KG=K                                               'and pointer to it
3370      COLOR 0,7                                          'display side comment
3380       LOCATE 2*K-1,42
3390       PRINT " largest in current pass so far"
3400      Z$=INPUT$(1)
3410      NEXT K
3500 REM * largest item in current cycle has been found *
3510      COLOR 0,7                                          'display side comment
3520      LOCATE 2*(M-J)+1,42
3530       IF KG<M-J+1 THEN 3560
3540       PRINT "in place already" STRING$(15,219)
3550        GOTO 3580
3560       PRINT "swap" STRING$(20,219)
3570      SWAP X(KG),X(M-J+1)                                              'swap
3580      Z$=INPUT$(1)
3590      Q=M-J+1: GOSUB 5300            'mark position with hyphens
3600      Q=KG: GOSUB 5100               'redisplay in new position
3610       GOSUB 5200                             'erase side comment
3620      Q=M-J+1: GOSUB 5100            'redisplay other swapped item
3630       GOSUB 5200
3640      Z$=INPUT$(1)
3650     ·Q=KG: GOSUB 5000                                    'de-highlight
3660      Q=M-J+1: GOSUB 5000
3670       Z$=INPUT$(1)
3680     NEXT J
3690     GOTO 4000
3999 :
4000 REM * sort is complete *
4010     LOCATE 2,32                         'erase the last line of hyphens
4020      PRINT SPC(6)
4030     LOCATE 24,1
4040      PRINT SPC(79)                            'clear the instruction line
4050     LOCATE 24,20
4060      PRINT "Press A for another example, Q to quit: ";            'prompt
4070      ON INT(INSTR("AaQq",INPUT$(1))/2) GOTO 200,7000              'branch
4080       GOTO 200
4999 :
5000 REM * display an entry *
5010     COLOR 7,0                                          'white on black
5020      LOCATE 2*Q-1,32                                   'vertical position
5030       PRINT USING "######";X(Q);                'display right aligned
5040      RETURN
5100 REM * highlight an entry *
5110     COLOR 0,7                                          'black on white
5120      LOCATE 2*Q-1,32
5130       PRINT USING "######";X(Q);                'display right aligned
5140     RETURN
5200 REM * erase a side comment *
```

```
521Ø   COLOR 7,Ø
522Ø     LOCATE 2*Q-1,42                          'cursor at start position
523Ø       PRINT TAB(79)                            'erase to end of line
524Ø   RETURN
53ØØ REM * draw hyphens *
531Ø   COLOR 7,Ø
532Ø     LOCATE 2*Q,32              'erase line of hyphens from previous cycle
533Ø       PRINT SPC(6)
534Ø     LOCATE 2*Q-2,32            'rule a line of hyphens above sorted items
535Ø       PRINT "------"
536Ø   RETURN
54ØØ REM * erase an entry and side comment *
541Ø   COLOR 7,Ø
542Ø     LOCATE 2*Q-1,32                          'cursor at start position
543Ø       PRINT TAB(79)                            'erase to end of line
544Ø   RETURN
5999 :
6ØØØ REM * error trap *
6Ø1Ø   COLOR 7,Ø                                       'default colors
6Ø2Ø     LOCATE 25,1                                      'diagnostic
6Ø3Ø       PRINT "Error" ERR "in line" ERL "- press space bar to continue";
6Ø4Ø   Z$=INPUT$(1)
6Ø5Ø   RESUME 7ØØØ
6999 :
7ØØØ REM * quit *
7Ø1Ø   LOCATE 1,1                       'position the cursor for the Ok message
7Ø2Ø     ON ERROR GOTO Ø                            'reset error trap
7Ø3Ø   END
```

The control structure of the insertion sort section of SIMPSORT is dominated by a FOR . . . NEXT loop. Successive cycles of this loop focus attention on successive elements of the original list:

```
2030 FOR J=2 TO M
2070   XC=X(J)
. . .
2710 NEXT J
```

Within the outer loop, the correct position of XC relative to the earlier elements is found by an inner WHILE . . . WEND loop. This slides the elements that are larger than XC forward in the array.

```
2310   K=J
2410   WHILE XC<X(K-1)  AND  K>1
. . .
2470     X(K)=X(K-1)
. . .
2510     K=K-1
2560   WEND
```

Following the loop, XC is slid into the appropriate slot.

```
2650   X(K)=XC
```

The control structure of the selection sort section is dominated by a FOR . . . NEXT loop, too. The first cycle of this loop ensures that the largest element is in the final slot of the array. The second cycle ensures that the next largest element is in the next to final slot, and so on.

```
3030   FOR J=1 TO M-1
. . .
3680   NEXT J
```

Within this loop, KG and XG are initialized to zero. An inner FOR . . . NEXT loop then finds the index and value of the largest of the elements in the first J slots of the array. When two or more elements tie for this largest value, the one with the highest index is chosen.

```
3110   FOR K=1 TO M-J+1
. . .
3150     IF X(K)>=XG THEN . . .
. . .
3350       XG=X(K)
3360       KG=K
. . .
3410   NEXT K
```

The largest element found in this loop is swapped with the element in the position that it must occupy, unless it is already in place.

```
3530   IF KG<M-J+1 THEN . . .
. . .
3570     SWAP X(KG),X(M-J+1)
```

Suggested Activities

1. Expand the program SIMPSORT:
    (*a*). To make it generate numbers that are negative as well as positive when the random number option is used
    (*b*). To let it sort numbers that are more than five characters long, for example, integers that are outside the range -9999 to 99999, numbers that contain digits following a decimal point, or numbers expressed in scientific notation, or hexadecimal numbers
    (*c*). To let it sort more than 10 numbers that are displayed in several columns
    (*d*). To make the random number option generate different sequences of numbers whenever the program is executed
    (*e*). To give the user the option of seeing the sort proceed automatically or in response to successive key presses

(f). To use a list that can be displayed across a single line of the screen (perhaps in width 80) while retaining the results of successive actions (either individual transpositions or complete sweeps) on successive lines

(g). To print the history of a sort in which the points of successive actions are flagged, for example, by words, asterisks, or underlining

2. Alter the program (a) to sort numbers into decreasing numerical sequence and (b) to let the user specify whether an individual sort is to be into the sequence that increases or decreases.

3. Streamline the bubble sort algorithm in the following ways. Use suitable display tactics to mark what is being done.

(a). Terminate the sort after the first sweep (that is, the first cycle) through the outer loop that causes no transpositions.

(b). In each sweep, retain the position of the very first slot to be used in a transposition. Start the next sweep from the preceding slot. This avoids reworking the initial portion of the list that is in order.

(c). In each sweep, retain the position of the very last slot to be used in a transposition. End the next sweep at this point if a transposition is not made into the slot immediately before it. This avoids reworking a final portion of the list that is in order.

4. Modify the selection sort algorithm to start each outer cycle by comparing each item in turn with its successor until either (a) two successive items that are out of order are found or (b) the end of the sweep is reached without finding any items out of order. In case (a) use the larger item as the largest item so far and proceed through the rest of the sweep in the manner already described. In case (b) signal that the list is sorted.

5. Adapt the program SIMPSORT to sort a collection of character strings:

(a). Into alphabetic order

(b). Into reverse alphabetic order

(c). In order of increasing length with each group of strings of uniform length arranged alphabetically

6. Write a program that (a) lets you type, for each of 50 or fewer people, the name and social security number, (b) lets you sort the data by using the name as the key, and (c) lets you sort the data by using the social security number as the key.

7. Extend Activity 6 to let you save the sorted files on disk.

1.2   Seeing Numbers Move

The program BUBLPLUS adds a further degree of animation to the demonstration of the bubble sort process. It begins, in the same way as the bubble option of SIMPSORT, by displaying a list of numbers that you can type or make the program generate. Then, when two numbers are swapped, you see them stay highlighted in their starting positions while copies move outward, pass each other, and move into their new positions along two smooth paths. The number from the upper slot moves out half as far as the number from the lower slot. The larger number then moves down while the smaller number moves up. The two numbers then move in again. The moving numbers are displayed in reverse video. When they cross the word "swap," the letters shine through the numerals.

Program Variables

BUBLPLUS uses most of the variables that are used by SIMPSORT and also the following:

U%	The array in which the bit map of a completely blank portion of the screen, 33 pixels wide and 10 pixels deep, is stored
S%	The array that stores the bit map of the 33 by 10 pixel portion of the screen containing the larger number of a pair that is being swapped
T%	The corresponding array for the smaller number
X1,Y1	The coordinates of the top-left portion of the screen that contains the moving copy of the larger number being swapped
X2,Y2	The corresponding coordinates for the smaller number
X,Y1S	The coordinates of the initial position of the larger number
X,Y2S	The coordinates of the initial position of the smaller number
T	The indexes of the loops that control the movement

Program Notes

The program works like the bubble option of SIMPSORT with a few additional features. The RANDOMIZE TIMER statement is active and initiates a different sequence of random numbers each time the program is run. The SCREEN 2 statement switches to high-resolution graphics at the start of the program to allow the "dancing numbers" to be produced by GET and PUT statements. The action of the GET statement is illustrated by

160 GET (0, 0) – (32, 9) , U%

This stores, in the array U%, the bit map of the picture that is on the rectangular portion of the screen with top-left coordinates (0,0) and bottom-right coordinates (32,9). The terms "source area" and "source picture" will be used for the area referenced by a GET and for the picture that the GET stores, respectively. The particular picture stored by statement 160 is completely blank because the preceding statement cleared the screen. The statements

```
2140 GET  (X,Y1)-(X+32,Y1+9),S%
2150 GET  (X,Y2)-(X+32,Y2+9),T%
```

store the pictures of the numbers being compared in the current cycle of the sorting process.

A PUT statement redisplays the picture that was stored by a GET. The general form is

```
PUT (x,y),u,option
```

where (x,y) are the coordinates of the top-left corner of the *target area* on which the picture is redisplayed and u stands for the name of the array in which the picture was stored. To explain the effects of the different options, it is convenient to use the terms "undercoat picture" for what is on the target area before the PUT statement is executed and "overcoat picture" for what is being redisplayed. The option PSET displays the overcoat picture without regard to the undercoat. The option PRESET displays the reverse video of the overcoat. Consequently, the statements

```
2160 PUT  (X,Y1),U%,PRESET
2170 PUT  (X,Y2),U%,PRESET
```

"white out" the 33 by 10 pixel target areas containing the two numbers that are highlighted in the current cycle. The XOR option, however, acts as follows

Pixel in undercoat	Off	Off	On	On
Pixel in overcoat	Off	On	Off	On
Pixel afterwards	Off	On	On	Off

The statements

```
2180 PUT  (X,Y1),S%,XOR
2190 PUT  (X,Y2),T%,XOR
```

therefore take the pictures of the numbers in S% and T% and display them in reverse video, because all the pixels in the two target areas are on as the result of statements 2160 and 2170.

When the undercoat is completely blank, the XOR option displays the overcoat picture in normal video. Furthermore, two applications of the same overcoat picture using the XOR option restore the target area of the screen to the original undercoat picture, whatever it was at the outset. This is very important in animation.

Pixel in undercoat	Off	Off	On	On
Pixel in overcoat	Off	On	Off	On
Pixel after 1st PUT . . . XOR	Off	On	On	Off
Pixel after 2nd PUT . . . XOR	Off	Off	On	On

The further details of the dancing numbers effect are simple. The pictures of the numbers that are now in reverse video are stored in S% and T% by two more GET statements that refer to the same source areas as before.

```
2200 GET (X, Y1) - (X+32, Y1+9) , S%
2210 GET (X, Y2) - (X+32, Y2+9) , T%
```

The subroutine that swaps two numbers on the screen begins with a loop that displays the first number in reverse video, displaced 16, 32, 48, 64, and 80 pixels to the right successively. The second number is displaced 32, 64, 96, 128, and 160 pixels. A second FOR . . . NEXT loop moves the larger number down and the smaller number up by changing the coordinates Y1 and Y2 in opposite directions and jumping to the display-and-erase subroutine. The two numbers are then moved into their new slots by another FOR . . . NEXT loop that is analogous to the loop that moved them out, but with X1 and X2 decreasing this time.

The display and erase subroutine simply uses

1. A pair of PUT . . . XOR statements to display the two numbers moved sideways from the list
2. A short delay loop
3. Another pair of PUT . . . XOR statements to reverse the effect of the first pair

This tactic allows the word "swap" to shine through when a number is displayed on top of it and restores the word "swap" when the number moves on.

```
100 REM * BUBLPLUS demonstrates bubble sort with animated swapping *
101 :
110   ON ERROR GOTO 3000                                'set error trap
120   SCREEN 2                                  'high resolution graphics
130   DIM X(10)                          'array of numbers to be sorted
140   DIM S%(26),T%(26),U%(26)            'arrays to store bit maps
150   KEY OFF: CLS                               'clear the screen
160   GET (0,0)-(32,9),U%               'store bit map of blank strip
199 :
200 REM * start a demonstration *
210   CLS                                         'clear the screen
220   LOCATE 23,27                                      'caption
230   PRINT "Bubble sort demonstration";
240   LOCATE 25,9                                             'prompt
250   PRINT "Press R for a list of random numbers, T to type your own data:";
260   R$=INPUT$(1)                                 'store the response
270   CLS                                          'clear the screen
280   IF R$="T" OR R$="t" THEN 300 ELSE 600     'branch to use random numbers
299 :
300 REM * user provides the numbers *
310   LOCATE 24,16
320   PRINT "Please type 10 integers between -99 and 999";         'prompt
330   FOR I=1 TO 10
340     LOCATE 2*I+1,37
350     INPUT "",X(I)                              'store the ith number
360     X(I)=INT(X(I))                    'truncate if not an integer
370     IF -99<=X(I) AND X(I)<=999 THEN 450       'check it is in range
380       LOCATE 2*I+2,6
390       PRINT "Please stay between -99 and 999. ";         'diagnostic
400       PRINT "Press space bar and try again."
410       Z$=INPUT$(1)
420       LOCATE 2*I+1,1
430       PRINT SPACE$(160);                        'erase the diagnostic
440     GOTO 340
450     LOCATE 2*I+1,37
460     PRINT SPC(40)                'clear the line on which it was typed
470     Q=I: GOSUB 2000                   'redisplay it right aligned
480   NEXT I
490   LOCATE 24,1                         'clear the request for input
500   PRINT SPC(79)
510   GOTO 1000
599 :
600 REM * program generates random numbers *
610   RANDOMIZE TIMER                          'different sequence each time
620   FOR I=1 TO 10
630     X(I)=1+INT(999*RND)                   'number between 1 and 999
640     Q=I: GOSUB 2000                                'display it
650     FOR D=1 TO 100: NEXT D                             'pause
660   NEXT I
999 :
1000 REM * sort the numbers *
1010  LOCATE 24,15                                             'prompt
1020  PRINT "Press the space bar to proceed from step to step.";
1030  FOR J=10 TO 2 STEP -1                                'outer loop
1040    FOR K=1 TO J-1                                     'inner loop
1099 :
1100 REM * highlight the entries to be compared *
1110    GOSUB 2100                          'highlight the two numbers
1120    LOCATE 2*K+1,42
1130      IF X(K)<=X(K+1) THEN PRINT "do not ";  'display do not swap or swap
```

```
1140        PRINT "swap";
1150      Z$=INPUT$(1)
1199 :
1200 REM * swap if necessary *
1210      IF X(K)<=X(K+1) THEN 1240            'jump if items in order
1220        SWAP X(K),X(K+1)                   'swap them otherwise
1230        GOSUB 2310
1240          LOCATE 2*K+1,42                  'clear swap message
1250          PRINT SPC(12)
1260        PUT (X1,Y1),S%,PRESET        'de-highlight the first number
1270        PUT (X2,Y2),T%,PRESET                   'and the second
1310      NEXT K
1399 :
1400 REM * end an outer loop *
1410      IF J=10 THEN 1440
1420        LOCATE 2*J+2,32        'erase line of hyphens from previous cycle
1430        PRINT SPC(6)
1440        LOCATE 2*J,32          'rule a line of hyphens above sorted items
1450        PRINT "------"
1460      Z$=INPUT$(1)
1470    NEXT J
1499 :
1500 REM * sort is complete *
1510    LOCATE 4,32                         'erase the last line of hyphens
1520    PRINT SPC(6)
1530    LOCATE 24,1
1540    PRINT SPC(79)                          'clear the instruction line
1550    LOCATE 24,20
1560    PRINT "Press A for another example, Q to quit: ";        'prompt
1570    R$=INPUT$(1)
1580    IF R$="Q" OR R$="q" THEN 5000 ELSE 200          'quit or another
1999 :
2000 REM * display an element *
2010    LOCATE 2*Q+1,32
2030      PRINT USING "######"+SPACE$(12);X(Q)         'delete the swap message
2040    RETURN
2100 REM * highlight two entries *
2110    X=263                         'screen coord for character position 35
2120    X1=X: X2=X               'initial values for use in swapping movement
2130    Y1=16*K-1: Y2=16*K+15          'corresponding vertical coordinates
2140    GET (X,Y1)-(X+32,Y1+9),S%             'store bit map of first number
2150    GET (X,Y2)-(X+32,Y2+9),T%                   'and of second number
2160    PUT (X,Y1),U%,PRESET                    'blank out first number
2170    PUT (X,Y2),U%,PRESET                          'and second
2180    PUT (X,Y1),S%,XOR       're-display first number with colors reversed
2190    PUT (X,Y2),T%,XOR                         'likewise for second
2200    GET (X,Y1)-(X+32,Y1+9),S%   'save bit map of 1st number in reverse video
2210    GET (X,Y2)-(X+32,Y2+9),T%                  'likewise for second
2240    RETURN
2299 :
2300 REM * swap two numbers on the screen *
2305 REM                                  move the numbers to the left
2310    FOR T=0 TO 80 STEP 16
2320      X1=X+T                      'x coord to move 1st number to the left
2330      X2=X+2*T                    'x coord to move 2nd number twice as far
2340      GOSUB 2600                      'display in new positions briefly
2350    NEXT T
2360 REM                                 switch the numbers vertically
2370    Y1S=Y1                          'starting values of y coords
2380    Y2S=Y2
```

```
2390   FOR T=0 TO 16 STEP 2
2400      Y1=Y1S+T                            'y coord to move 1st number down
2410      Y2=Y2S-T                            'y coord to move 2nd number up
2420      GOSUB 2600                          'display in new positions briefly
2430   NEXT T
2435 REM                              move the numbers back to the right
2440      X1S=X1                             'starting values of x coords
2450      X2S=X2
2460   FOR T=0 TO 80 STEP 16
2470      X1=X1S-T                   'x coord to move 1st number to the left
2480      X2=X2S-2*T                 'x coord to move 2nd number twice as far
2490      GOSUB 2600                          'display in new positions briefly
2500   NEXT T
2510   RETURN
2599 :
2600 REM * move the numbers on the screen *
2610   PUT (X1,Y1),S%,XOR               'display 1st number in new position
2620   PUT (X2,Y2),T%,XOR                      'and 2nd number too
2630     FOR D=1 TO 50: NEXT D                        'pause
2640   PUT (X1,Y1),S%,XOR          'restore area just occupied by 1st number
2650   PUT (X2,Y2),T%,XOR                       'and by 2nd number too
2660   RETURN
2999 :
3000 REM * error trap *
3010   COLOR 7,0                                   'default colors
3020   LOCATE 25,1
3030     PRINT "Error" ERR "in line" ERL TAB(79)        'diagnostic
3040   RESUME 5000                              'reset error flags
4999 :
5000 REM * quit *
5010   LOCATE 1,1                      'position the cursor for the Ok message
5020   ON ERROR GOTO 0                          'reset error address
5030   END                                              'quit
```

After two numbers have been compared and, if necessary, swapped, the main part of the program de-highlights them. S% and T% contain pictures of the numbers in reverse video, and the PRESET option forms the reverse video of them, that is, the normal video display.

```
1260 PUT (X1,Y1),S%,PRESET
1270 PUT (X2,Y2),T%,PRESET
```

Note that the GET and PUT statements use source and target areas that are 33 pixels wide and 10 pixels high, even though each character is displayed on an 8 by 8 raster. The extra pixels provide a frame around the reverse video numbers. Without the frame, some of the unlit pixels in the reverse video numbers would appear to "run into" the unlit background.

Suggested Activities

1. Adapt BUBLPLUS to demonstrate the transpositions in the operation of the selection sort by the same animation method.

2. Use the same technique to demonstrate the transpositions in a program that sorts strings by using the bubble sort or selection sort algorithm.

3. Adapt BUBLPLUS to medium-resolution graphics and compare the visual effect with the high-resolution version.

1.3 Sorting the Pointers

Computers can swap two numbers quite quickly, but swapping two strings takes longer, particularly if the strings are lengthy. The program PTRSORT shows how a list of strings can be put in order by sorting the numbers that specify the positions of the items in the unsorted list. For example, the alphabetic ordering of the list shown in (a) is expressed by the "pointer list" shown in (b).

dog	2
cat	5
gnu	1
elk	4
cow	10
rat	8
hog	3
fox	7
yak	6
ewe	9
(a)	(b)

The first item in the pointer list, 2, shows that the second item of the unsorted list, cat, is the first alphabetically. The second item in the pointer list, 5, shows that the fifth item of the unsorted list, cow, is the second alphabetically. Likewise, the third, fourth, . . . , tenth items of the alphabetically sorted list are the first, fourth, . . . , ninth items of the unsorted list.

The program PTRSORT constructs the sorted pointer list without moving a single string, and then uses it to display the list of strings in alphabetic order. PTRSORT begins by prompting you to press P for a preset list or T to type your own data. If you press P, the unsorted list of 10 three-letter animals appears at the left of the screen, with the numbers 1 to 10 in a second column (c). Then you are prompted to press the space bar to proceed to the next step. You respond, and the numbers 1 to 10 appear in a third column at the right of the screen (d).

dog	1		dog	1		1
cat	2		cat	2		2
. . .			. . .			
ewe	10		ewe	10		10
(c)			(d)			

The numbers in the third column are the pointers that will move. Press the space bar again. The first two pointers turn red. Lines shoot across to the items that they specify, and these entries turn red too. The numbers that are displayed in red on the screen are shown here by underlining (e). The items dog and cat are compared. Because they are out of alphabetic order, the word "swap" appears alongside the first pointer. Then the first two pointers are swapped, and the lines connecting them to their referents are redrawn (f).

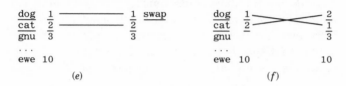

(e)   (f)

Press the space bar again. The top two lines of information change to the original color and the connecting lines and the word "swap" vanish. Then the pointers in the second and third lines turn red. Lines shoot across to their referents, which turn red, too (g). The items dog and gnu are compared. They are in alphabetic order, so the words "do not swap" appear (h).

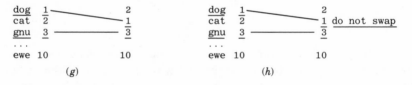

(g)   (h)

Go on pressing the space bar. The sequence of numbers in the pointer list after successive key presses is

```
2, 1, 3, 4, 5, 6, 7, 8, 9, 10     after comparing dog and gnu
2, 1, 4, 3, 5, 6, 7, 8, 9, 10     after comparing gnu and elk
2, 1, 4, 5, 3, 6, 7, 8, 9, 10     after comparing gnu and cow
2, 1, 4, 5, 3, 6, 7, 8, 9, 10     after comparing gnu and rat
2, 1, 4, 5, 3, 7, 6, 8, 9, 10     after comparing rat and hog
2, 1, 4, 5, 3, 7, 8, 6, 9, 10     after comparing rat and fox
2, 1, 4, 5, 3, 7, 8, 6, 9, 10     after comparing rat and yak
2, 1, 4, 5, 3, 7, 8, 6, 10, 9     after comparing yak and ewe
```

The pointer 9 to yak, the item that comes last alphabetically, is now in position 10, the final position of the pointer list. A line of hyphens is displayed above it. In each cycle, pointers and referents change color, lines are drawn and erased, and the words "swap" and "do not swap" are displayed and erased appropriately.

Pressing the space bar again moves back to the start of the pointer list. The first two items turn red, and they are joined to their referents. These turn red also (*i*). The message "do not swap" appears, and attention moves on to the items identified by the pointers in the second and third lines after you press the space bar (*j*).

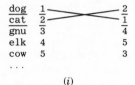 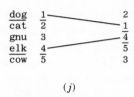

(*i*)   (*j*)

The word "dog" precedes elk alphabetically, so "do not swap" appears and attention moves on after you press the space bar (*k*).

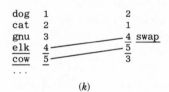

(*k*)

Go on pressing the space bar. The sequence of pointers after successive key presses is

```
2, 1, 5, 4, 3, 7, 8, 6, 10, 9    after comparing elk and cow
2, 1, 5, 4, 3, 7, 8, 6, 10, 9    after comparing elk and gnu
2, 1, 5, 4, 3, 7, 8, 6, 10, 9    after comparing gnu and hog
2, 1, 5, 4, 3, 8, 7, 6, 10, 9    after comparing hog and fox
2, 1, 5, 4, 3, 8, 7, 6, 10, 9    after comparing hog and rat
2, 1, 5, 4, 3, 8, 7, 10, 6, 9    after comparing rat and ewe
```

At this juncture, the pointers in the last two positions of the pointer list refer inevitably to the two animals that come last in alphabetical order.

The next sweep compares, in the corresponding manner, cat (item 2) and dog (item 1) and leaves the pointers unchanged, dog (item 1) and cow (item 5) and swaps the pointers, dog (item 1) and elk (item 4) and leaves the pointers unchanged, elk (item 4) and gnu (item 3) and does not swap, gnu again and fox (item 8) and swaps, gnu again and hog (item 7) but does not swap, and hog (item 7) and ewe (item 10) and swaps, giving the pointer sequence:

```
2, 5, 1, 4, 8, 3, 10, 7, 6, 9
```

The last three items in the pointer list now correspond to the three animals that come last alphabetically. The next two sweeps change the pointer list, respectively, to

2, 5, 1, 4, 8, 10, 3, 7, 6, 9   and   2, 5, 1, 4, 10, 8, 3, 7, 6, 9

This ensures that the last five pointers are in place. The program goes through another four sweeps, which end at the fifth, fourth, third, and second positions in the pointer list, respectively, without making any further changes. The sort is then declared final. Another key press makes the program cycle through the pointer list and displays the referents alongside it in alphabetic order. You are then prompted to type A for another example or Q to quit. If you type A, the prompt to press P for a preset list or T for your own data reappears. Pressing P repeats the example that has been described. Pressing T leads to the prompt to type 10 three-letter words. The program positions the cursor in the middle of successive lines of the screen for the successive entries. When you have typed the tenth, the pointers appear alongside their referents and you are prompted to press the space bar to cycle through the process of reordering the pointers. If, during the input stage, you enter a string that is more than three characters long, the string is truncated.

Program Variables

S$(i)   The ith string in the list

P(q)   The pointer currently in the qth position of the pointer list

R$ The response to prompts for one-letter options

I   The index of an element of the S$ array

Q   The index of the pointer list

J,K   The indexes of the outer and inner loops of the sort process

Program Notes

The bubble sort algorithm is embodied in the nested loop that is indexed by J and K, just like the corresponding portions of the programs SIMPSORT and BUBLPLUS. The index K identifies items in the pointer list P. The ordering criterion is applied to the elements of S$ to which these pointers refer. The comparison statement uses subscripted subscripts (indirect addressing).

```
1210  IF  S$(P(K))<=S$(P(K+1))  THEN  1310
```

```
100 REM * PTRSORT demonstrates pointer sort based on bubble algorithm *
101 :
110  ON ERROR GOTO 3000                            'set the error trap
120  SCREEN 1: KEY OFF: COLOR 1,0                   'set up the screen
130  DIM S$(10),P(10)
140  DATA dog,cat,gnu,elk,cow,rat,hog,fox,yak,ewe
199 :
200 REM * start a demonstration *
210  CLS                                            'clear the screen
220  LOCATE 21,5                                    'caption
230   PRINT "Pointer sort demonstration";
240  LOCATE 23,5                                    'prompt
250   PRINT "Press P for a pre-set list,";
260  LOCATE 25,5                                    'prompt
270   PRINT "     T to type your own data:";
280  R$=INPUT$(1)                                   'store the response
290  CLS                                            'clear the screen
300  IF R$="T" OR R$="t" THEN 400 ELSE 600          'branch to use preset list
399 :
400 REM * user provides the data *
410  LOCATE 24,1                                    'prompt for data
420   PRINT "Please type 10 three-letter words";    'to be sorted
430   FOR I=1 TO 10
440    LOCATE 2*I-1,1
450    INPUT; "",S$(I)                              'store the ith word
460   IF LEN(S$(I))<=3 THEN 520                     'jump if short enough
470    S$(I)=LEFT$(S$(I),3)                         'truncate otherwise
480    LOCATE 2*I-1,4
490     PRINT TAB(40)
500    LOCATE 2*I-1,4
520   PRINT USING "###";I                           'display the slot number
530   P(I)=I                                        'set the pointer
540   NEXT I
550   LOCATE 24,1                                   'clear the request for input
560    PRINT SPC(79)
570    GOTO 1000                                    'go on to sort
599 :
600 REM * program uses the pre-set list *
610   FOR I=1 TO 10
620    READ S$(I)                                   'transfer word from DATA statement
630    P(I)=I                                       'set the pointer
640    LOCATE 2*I-1,1
650     PRINT USING "\ \ ##";S$(I),I                'display the word and its slot #
660   NEXT I
670   RESTORE
999 :
1000 REM * sort the list *
1010  LOCATE 24,1                                   'prompt
1020   PRINT "Press space bar for next step";
1030    GOSUB 2400                                  'wait for key press
1040    FOR Q=1 TO 10
1050     LOCATE 2*Q-1,19
1060     PRINT USING "##";P(Q)                      'display the qth pointer
1070    NEXT Q
1080   FOR J=10 TO 2 STEP -1                        'outer loop
1090    FOR K=1 TO J-1                              'inner loop
1099 :
1100 REM * color the entries to be compared and the pointers *
1110    POKE &H4E,2                                 'switch to red as foreground color
1120     Q=K: GOSUB 2000                            'emphasize item with Kth pointer
```

```
1130     Q=K+1: GOSUB 2000                       'emphasize item with (K+1)th pointer
1140     GOSUB 2400                                    'wait for key press
1150     LOCATE 2*K-1,22
1160     IF S$(P(K))<=S$(P(K+1)) THEN PRINT "do not ";        'display comment
1170       PRINT "swap";
1180     GOSUB 2400                                    'wait for key press
1199 :
1200 REM * swap if necessary *
1210   IF S$(P(K))<=S$(P(K+1)) THEN 1310              'jump if items in order
1220   Q=K                                    'display referant of kth pointer
1230     GOSUB 2200
1240   Q=K+1                                'display referant of (k+1)th pointer
1250     GOSUB 2200
1260   SWAP P(K),P(K+1)                                 'swap pointers
1270   Q=K                              'display referant of updated kth pointer
1280     GOSUB 2000
1290   Q=K+1                        'display referant of updated (k+1)th pointer
1300     GOSUB 2000
1310   POKE &H4E,3                          'switch back to white foreground
1320   LOCATE 2*K-1,22                               'clear the swap message
1330     PRINT SPC(12)
1340     GOSUB 2400                                    'wait for key press
1350   Q=K                          'de-emphasize the referant of the kth pointer
1360       GOSUB 2000
1370     GOSUB 2200                                    'wait for key press
1380   Q=K+1                      'de-emphasize the referant of the (k+1)th pointer
1390     GOSUB 2000
1400     GOSUB 2200                                    'wait for key press
1410   NEXT K
1499 :
1500 REM * end an outer loop *
1510   IF J=10 THEN 1540
1520     LOCATE 2*J,19                'erase line of hyphens from previous cycle
1530     PRINT SPC(2)
1540     LOCATE 2*J-2,19              'rule a line of hyphens above sorted items
1550       PRINT "--"
1560     LOCATE 2*J-1,23
1570       PRINT S$(P(J))            'de-emphasize the referant of the jth pointer
1580     GOSUB 2400                                    'wait for key press
1590   NEXT J
1699 :
1700 REM * sort is complete *
1710   LOCATE 2,19                              'erase the last line of hyphens
1720     PRINT SPC(2)
1730   LOCATE 1,23
1740     PRINT S$(P(1))            'de-emphasize the referant of the 1st pointer
1750   LOCATE 24,1
1760     PRINT SPC(79)                           'clear the instruction line
1770   LOCATE 24,1
1780     PRINT "(A)nother example or (Q)uit: ";                 'prompt
1790     R$=INPUT$(1)
1800     IF R$="Q" OR R$="q" THEN 4000 ELSE 200            'quit or another
1999 :
2000 REM * display an element *
2010   R=P(Q)                               'get actual slot # from pointer
2020   LOCATE 2*R-1,1              'locate appropriate line of original list
2030     PRINT USING "\ \ ##";S$(R),R          're-display item and slot #
2040   LOCATE 2*Q-1,19                      'locate line in pointer list
2050     PRINT USING "##";R                   're-display the pointer
2060   LINE (50,16*R-12)-(142,16*Q-12),2          'join pointer to referant
```

```
2070   RETURN
2199 :
2200 REM * erase a line from pointer to referant *
2210   R=P(Q)                                              'get slot # from pointer
2220   LINE (50,16*R-12)-(142,16*Q-12),0                        'erase the line
2230   RETURN
2399 :
2400 REM * wait for key press *
2410   Z$=INPUT$(1)
2420    IF ASC(R$)=27 THEN 4000                                'branch if Escape
2430    RETURN
2999 :
3000 REM * error trap *
3010   LOCATE 21,1
3020   PRINT "Error" ERR "in line" ERL;                               'diagnostic
3030   POKE &H4E,3                                          'ensure default color
3040   RESUME 4000                                              'clear error flags
3999 :
4000 REM * quit *
4010   LOCATE 21,1                                'position the cursor for the Ok message
4020   LOCATE 21,1                                'position the cursor for the Ok message
4030   ON ERROR GOTO 0                                      'reset error trap address
4040   END                                                              'quit
```

When the referents are out of order, the pointers are swapped:

```
1260 SWAP P(K),P(K+1)
```

Suggested Activities

1. Adapt and extend the PTRSORT program:
   (a). To print the history of a pointer sort
   (b). To store the list of strings on disk in alphabetic order at the end of a sort
   (c). To display the list of strings in reverse alphabetic order at the end of the sort

2. Write a program analogous to PTRSORT that prompts for a list of strings, sorts the pointers into the sequence that corresponds to alphabetic order of the strings, and prints the strings in alphabetic order based on:
   (a). The insertion sort algorithm
   (b). The selection sort algorithm

3. Write a program analogous to PTRSORT that works in the text mode and uses graphics characters to form the lines that join pointers to their referents. (*Hint:* See the linked list program in Chapter 3.)

1.4   A Quicker Way to Sort

The program QWIKSORT demonstrates the *quicksort* algorithm that is faster for long files than the elementary sorting algorithms of Section 1.1. The program prompts you to type how many numbers you want it to sort. The limit in the program is 25. Then it prompts you to press R for random numbers or T to type the numbers. Suppose you entered 7 and typed R in response to the prompts. The program displays

        71 68 48 99 65 14 37

at the top of the screen.

The program begins the sorting operation by selecting the rightmost element as the first *pivot*. The general strategy is to find the *target position* that the pivot would occupy if the list were fully sorted and to put the pivot there. In the course of doing so, the program moves any elements that are larger than the pivot from the left to the right of the target position. The program also moves any smaller elements from the right to the left of that position. The movements are made by swapping pairs of elements in the manner that we are about to explain.

You make the program proceed from step to step by pressing the space bar. The process begins by highlighting the pivot and comparing it with the leftmost element, which is 71 in the present example. The highlighting is shown here by boldface. An arrow (->) is displayed underneath the element that is being compared.

        [71 68 48 99 65 14 **37**]
        ->

The program finds that the marked element is larger than the pivot; therefore, it will move the marked element a little later. The program now starts to work from the right of the list to find if there is an element that is smaller than the pivot. The search starts with the element on the left of the pivot, 14. The 14 is marked by an arrow.

        [71 68 48 99 65 14 **37**]
        ->   <-

The element 14 is smaller than the pivot, so it is swapped with the 71. Both elements turn blue, to show that they are correctly placed with respect to the target position. The list is displayed in its new arrangement below the original list.

        [14 68 48 99 65 71 **37**]

A line of hyphens is drawn below this record of the first round of activity. The list is repeated in its rearranged form for you to follow the next round. Another left-to-right inspection for an element that is larger than the pivot now starts at the element on the right of the 14, which is 68. The complete display at this point is:

```
[71  68  48  99  65  14  37]
 ->                  <-
[14  68  48  99  65  71  37]
----------------------
[14  68  48  99  65  71  37]
    ->
```

The element 68 also is larger than the pivot. Therefore, it will need to be moved, too. Another right-to-left inspection for an element that is smaller than the pivot starts at the 65.

```
[14  68  48  99  65  71  37]
    ->          <-
```

This element is larger than 37. It turns blue, to show that it is on the correct side of the target position. Then the element on its left, 99, is examined. The arrow moves under the 99 accordingly:

```
[14  68  48  99  65  71  37]
    ->      <- <-
```

This element also is larger than the pivot. It turns blue. The arrow moves left again to show that the element 48 is being examined.

```
[14  68  48  99  65  71  37]
    -> <- <- <-
```

This element also is larger than the pivot. It turns blue, and the program moves to the next element on the left, 68.

The 68, however, had been marked with an arrow for movement to the right as a result of the actions described above. It follows that the only elements between 68 and the pivot must be larger than the pivot. Also, as a result of the earlier actions, 68 is certain to be at the right of any elements that are smaller than the pivot. The arrow under the 68 is changed to <> to show that the pointers have collided. As a consequence, 68 will be swapped with the pivot.

```
[14  68  48  99  65  71  37]
    <> <- <- <-
```

The program swaps 37 and 68. This moves the 37 into the position that is certain to be on the right of all the elements that are smaller and on the left of all the elements that are larger. Thus, the 37 is in its target position. The program turns 37 red to show that it is in place. We show this here by underlining.

```
[14 37 48 99 65 71 68]
```

A line of = symbols is displayed to rule off the activity that has brought an item into place.

The problem of sorting the original list has been reduced to sorting the elements on the left of the pivot in its new position and then sorting the elements on the right. Accordingly, the data is repeated on the next line of the screen.

```
14 37 48 99 65 71 68
```

When you press the space bar again, the brackets are placed around the leftmost sublist that contains elements which are not yet known to be in place. The rightmost element in this sublist is highlighted.

```
[14] 37 48 99 65 71 68
```

The sublist under current attention contains only one element 14. Consequently, the program recognizes that this element is in its final position. The 14 turns red. A line of = symbols is drawn, and the data is repeated in its new order lower down the screen. The brackets are moved around the next sublist that remains to be sorted. The right-most element is highlighted to show that it is the pivot of the sublist under attention. An arrow is put under the leftmost element to show it is being compared with the pivot.

```
14 37 [48 99 65 71 68]
      ->
```

The element 48 is less than 68. Therefore, it turns blue and the arrow moves to show that the left-to-right inspection has reached the next element, 99.

```
14 37 [48 99 65 71 68]
      -> ->
```

This element is larger than the pivot. Accordingly, a right-to-left search is started at the 71 for an element that is smaller than the pivot.

```
14 37  [48 99 65 71 68]
        -> ->     <-
```

The element 71 is larger than the pivot. Consequently, it turns blue and the right-to-left search moves to the element on its left, 65.

```
14 37  [48 99 65 71 68]
        -> -> <- <-
```

The 65 is smaller than the pivot. Consequently, it is swapped with the 99 and turns blue.

```
14 37  [48 65 99 71 68]
```

At this point we can be certain that the 65 is on the left of the target position. The program starts to search for an element that is larger than the pivot, beginning one place to the right of the 65. Because adjacent elements were just swapped, however, the search is started with the element 99 that, as a result of earlier action, is known to be larger than the pivot. The arrow appears under the 99, accordingly, and a bar | shows that this crossing has occurred.

```
14 37  [48 65 99 71 68]
               |->
```

Swapping the 99 with the pivot moves the pivot into the required target position. Accordingly, the pivot turns red.

```
14 37  [48 65 68 71 99]
```

This is ruled off with a line of = symbols, and the rearranged data is repeated.

```
14 37  48 65 68 71 99
```

The brackets close around the leftmost sequence of elements not yet known to be in their final positions. The rightmost of these is highlighted to show it is the new pivot.

```
14 37  [48 65] 68 71 99
        ->
```

```
1ØØ REM * QWIKSORT demonstrates quick sort (partition exchange) *
1Ø1 :
11Ø  ON ERROR GOTO 6ØØØ                                          'set error trap
12Ø  KEY OFF: SCREEN Ø: WIDTH 8Ø                                 'set up the screen
121 :
2ØØ REM * prompt *
21Ø   COLOR 7,Ø                                                  'default colors
22Ø   CLS
23Ø   GOSUB 55ØØ                                                       'prompt
24Ø   INPUT; "How many numbers do you want to sort (limit is 25): ",L
25Ø   IF 1<=L AND L<=25 THEN 3ØØ
26Ø     GOSUB 55ØØ                                               'invalid response
27Ø     PRINT L "is invalid - press space bar and try again";
28Ø     Z$=INPUT$(1)
29Ø     GOTO 23Ø
3ØØ   DIM X(L),C(L)                         'arrays for values and color codes
31Ø   DONE$=STRING$(L,"Ø")                  'string to monitor progress
32Ø   GOSUB 55ØØ                                                       'prompt
33Ø   PRINT "Press R for random numbers, T to type the numbers";
34Ø   R$=INPUT$(1)
35Ø   IF R$="T" OR R$="t" THEN 61Ø                          'branch for typed input
399 :
4ØØ REM * set up an array of random numbers *
41Ø   GOSUB 55ØØ
42Ø   PRINT "Please wait - computing"
43Ø   FOR I=1 TO L
44Ø   X(I)=1+INT(99*RND)                                    'form a random number
45Ø   FOR J=1 TO I-1                                     'form another if an earlier
46Ø    IF X(J)=X(I) THEN 44Ø                                  'value is repeated
47Ø   NEXT J
48Ø   NEXT I
49Ø   GOTO 8ØØ
599 :
6ØØ REM * input the numbers *
61Ø   FOR I=1 TO L
62Ø   CLS
63Ø    LOCATE 4,1
64Ø     PRINT "x(" MID$(STR$(I),2) "): ";                    'prompt for x(i)
65Ø     INPUT; "",X(I)
66Ø     IF Ø<X(I) AND X(I)<1ØØ AND X(I)=INT(X(I)) THEN 71Ø
67Ø      PRINT " - this is not an integer between 1 and 99."       'invalid
68Ø      PRINT "         press space bar and try again"
69Ø      Z$=INPUT$(1)
7ØØ      GOTO 62Ø                                            'allow retry
71Ø   NEXT I
799 :
8ØØ REM * display the list *
81Ø   CLS
82Ø   VCP=2                                            'vertical cursor position
83Ø   FOR I=1 TO L
84Ø    C(I)=7                                                'white foreground
85Ø   GOSUB 45ØØ                                        'display element in slot i
87Ø   NEXT I
999 :
1ØØØ REM * sort the list *
1Ø1Ø   WHILE INSTR(DONE$,"Ø")                     'cycle until all items in place
1Ø2Ø   J=INSTR(DONE$,"Ø")                      'left end of sublist to be sorted next
1Ø3Ø   K=INSTR(J,DONE$,"1")                           'right end of sublist
1Ø4Ø    IF K=Ø THEN K=L ELSE K=K-1                'allow for startup condition
1Ø5Ø   LB=3*J-2                                       'position of [
```

```
1Ø6Ø    RB=3*K+1                                          'position of ]
1Ø7Ø      GOSUB 5ØØØ
1Ø8Ø      GOSUB 48ØØ                             'display the [ and ]
1Ø9Ø      I=K                                    'prepare to highlight
11ØØ      C(I)=15                                'the pivotal element
111Ø      GOSUB 45ØØ                                  'redisplay it
1199 :
12ØØ REM * move pivot into place *
121Ø      M=J                          'pointer to leftmost item to be compared
122Ø      N=K-1                               'pointer to rightmost item
123Ø      IF M>N THEN 18ØØ
124Ø      WHILE M<=N+1                         'loop until pointers cross
125Ø      LOCATE VCP+1,3*M-2
126Ø      IF M<=N OR M=J THEN PRINT " ";
                      ELSE PRINT CHR$(179);    'display | if crossed already
127Ø      PRINT "->"                           'mark leftmost element
128Ø      GOSUB 5ØØØ                            'wait for key press
1299 :
13ØØ REM * move right through smaller elements
131Ø      WHILE X(M)<=X(K) AND M<=N             'loop through smaller elements
132Ø        GOSUB 41ØØ                          'turn mth item blue
133Ø        LOCATE VCP+1,3*M+1                  'move the marking arrow
134Ø        IF M<N THEN PRINT " ";
                      ELSE PRINT CHR$(179);     'display | if pointers crossed
135Ø        PRINT "->"
136Ø        GOSUB 5ØØØ                          'wait for key press
137Ø        M=M+1                               'proceed to next element
138Ø      WEND
139Ø      IF M<N THEN 15ØØ                      'proceed to search from right
14ØØ      IF X(M)>X(K) THEN SWAP X(M),X(K)      'move pivotal element
141Ø      GOTO 18ØØ
1499 :
15ØØ REM * move left through larger elements
151Ø      LOCATE VCP+1,3*N-1                    'mark the starting position
152Ø      PRINT "<-"
153Ø      GOSUB 5ØØØ
154Ø      WHILE X(N)>X(K) AND N>M               'loop through items larger than pivot
155Ø        GOSUB 42ØØ                          'redisplay nth item
156Ø        N=N-1                               'decrease the pointer
157Ø        IF N=M THEN 161Ø                    'jump if leftmost element passed
158Ø        LOCATE VCP+1,3*N-1
159Ø        PRINT "<-"                          'move the marking arrow
16ØØ        GOSUB 5ØØØ
161Ø      WEND
162Ø      IF M<N THEN 169Ø                      'smaller elements remain
163Ø        LOCATE VCP+1,3*N-1                  'move the marking arrow
164Ø        PRINT "<>"                          'pointers have crossed
165Ø        GOSUB 5ØØØ
166Ø      IF VCP>21 THEN GOSUB 35ØØ             'start new screen if needed
167Ø      SWAP X(M),X(K)                        'swap pivotal element into target slot
168Ø      GOTO 18ØØ
169Ø      GOSUB 3ØØØ                            'transpose smaller and larger element
17ØØ      M=M+1                                 'reset positions to start testing
171Ø      N=N-1                                 'for smaller and larger elements
172Ø      WEND
173Ø :
18ØØ REM * the original x(k) is in its final position *
181Ø      GOSUB 35ØØ
                                               'start new screen if needed
182Ø      C(M)=4                                'will turn pivot red
```

```
183Ø    IF M<>K THEN C(K)=3                          'and swapped element blue
184Ø   MID$(DONE$,M,1)="1"                           'slot j holds final item
185Ø    GOSUB 4ØØØ                                   'display the updated list
186Ø    GOSUB 48ØØ                       'mark next sublist to be processed
187Ø    LOCATE VCP+1,1
188Ø      PRINT STRING$(3*L,"=")                     'display line of = signs
189Ø    GOSUB 5ØØØ
19ØØ    FOR I=1 TO L
191Ø      IF C(I)=3 THEN C(I)=7                        'turn blue item white
192Ø    NEXT I
193Ø    GOSUB 4ØØØ                                   'display the updated list
194Ø   WEND
195Ø :
2ØØØ REM * prompt for further action *
2Ø1Ø   LOCATE 25,1
2Ø2Ø   PRINT "(S)ort another list or (Q)uit:";                       'prompt
2Ø3Ø   R$=INPUT$(1)
2Ø4Ø   IF R$="Q" OR R$="q" THEN 7ØØØ                 'jump if quit requested
2Ø5Ø   ERASE X,C                        'erase arrays to allow another demo
2Ø6Ø   GOTO 2ØØ                                    'go back to main prompt
2Ø7Ø :
3ØØØ REM * swap two elements *
3Ø1Ø   IF VCP>21 THEN GOSUB 35ØØ       'restart at top of screen if necessary
3Ø2Ø   SWAP X(M),X(N)                                                  'swap
3Ø3Ø   GOSUB 4ØØØ                                   'display the updated list
3Ø4Ø   GOSUB 41ØØ                                       'turn mth item blue
3Ø5Ø   GOSUB 42ØØ                                       'tunr nth item blue
3Ø6Ø   GOSUB 48ØØ                           'mark the sublist being sorted
3Ø7Ø   GOSUB 5ØØØ
3Ø8Ø   LOCATE VCP+1,1
3Ø9Ø      PRINT STRING$(3*L,"-")                     'display line of hyphens
31ØØ   GOSUB 4ØØØ                                   'display the updated list
311Ø   GOSUB 48ØØ                               'mark sublist being sorted
312Ø   LOCATE VCP+1,3*K-1                            'reposition the arrow
313Ø   RETURN
3499 :
35ØØ REM * start new screen if necessary *
351Ø   IF VCP<=21 THEN RETURN                        'new screen not needed
352Ø    ARL$=""
353Ø   FOR II=1 TO 79                      'preserve the line of markers
354Ø    ARL$=ARL$+CHR$(SCREEN(VCP+1,II))
355Ø   NEXT II
356Ø   CLS                                             'clear the screen
357Ø   VCP=Ø                               'reset the vertical position
358Ø   GOSUB 4ØØØ                               'display the revised list
359Ø   GOSUB 48ØØ                           'mark the ends of the sublist
36ØØ   LOCATE VCP+1,1
361Ø   PRINT ARL$                           'redisplay the line of markers
362Ø   GOSUB 5ØØØ
363Ø   RETURN
3999 :
4ØØØ REM * display the rearranged list *
4Ø1Ø   VCP=VCP+2                                   'vertical cursor position
4Ø2Ø   FOR I=1 TO L
4Ø3Ø    GOSUB 45ØØ                               'display the item in slot i
4Ø4Ø   NEXT I
4Ø5Ø   RETURN
41ØØ REM * show mth item is in place
411Ø    I=M
412Ø    C(M)=3                                         'turn item blue
```

```
4130    GOSUB 4500                                          'redisplay it
4140    RETURN
4200 REM * show nth item is in place
4210    I=N
4220    C(N)=3                                              'turn item blue
4230    GOSUB 4500                                          'redisplay it
4240    RETURN
4500 REM * display an item *
4510    COLOR C(I)
4520    LOCATE VCP,3*I-1                                    'position the item
4530    PRINT USING "##";X(I);                              'display it
4540    COLOR 7
4550    RETURN
4800 REM * show ends of sub list being sorted *
4810    LOCATE VCP,LB                                       'position the [
4820      PRINT "["                                         'display it
4830    LOCATE VCP,RB                                       'position the ]
4840      PRINT "]"                                         'display it
4850    RETURN
4999 :
5000 REM * wait for key press *
5010    Z$=INPUT$(1)
5020    IF R$=CHR$(27) THEN 7000                            'quit if Escape was pressed
5030    RETURN
5500 REM * clear prompt area *
5510    LOCATE 2,1
5520      PRINT SPC(79)                                     'clear prompt area
5530    LOCATE 2,1
5540    RETURN
5999 :
6000 REM * error trap *
6010    COLOR 7,0
6020    LOCATE 25,1
6030    PRINT "Error" ERR "in line " ERL TAB(79)            'diagnostic
6040    RESUME 7000                                         'clear error flags
6999 :
7000 REM * quit *
7010    LOCATE 1,1                                          'position the Ok message
7020    ON ERROR GOTO 0                                     'reset error trap address
7030    END
```

The 48 is smaller than the pivot. Moving the arrow to the right brings it under the pivot. A bar is used again to show that the –> has crossed into "forbidden territory."

```
14 37 [48 65] 68 71 99
        ->|->
```

This time, the excessive movement of the –> arrow to the right shows that the pivot is in place. It turns red. The rearranged data is repeated and the brackets close around what is now left of the sublist, which consists of elements that are not yet known to be in place.

```
14 37 [48] 65 68 71 99
```

This smaller sublist consists of just a single element; therefore, it must be in place. The rearranged data is repeated. The brackets close around the next sequence of elements not yet known to be in place.

<u>14</u> <u>37</u> <u>48</u> <u>65</u> <u>68</u> [71 **99**]
                        ->

This sublist is [71 99]. The pivot 99 is found to be in place by the same reasoning that found the 65 to be in place. The rearranged data is repeated and the brackets close around the 71.

<u>14</u> <u>37</u> <u>48</u> <u>65</u> <u>68</u> [**71**] <u>99</u>

The sublist now under attention contains just one element. Therefore it is in place. The program finds that all the elements are now known to be in place, and displays the data in its final arrangement, without brackets.

<u>14</u> <u>37</u> <u>48</u> <u>65</u> <u>68</u> <u>71</u> <u>99</u>

Consequently, the sort is complete.

One further detail concerning the display tactics must be mentioned. When the course of action reached the point depicted by

<u>14</u> <u>37</u> [48 **65**] <u>68</u> <u>71</u> <u>99</u>
          -> | ->

the screen was full. Consequently, when the space bar was pressed again, the screen cleared and this line was repeated at the top. The next key press then caused the 65 to turn red, and action continued in the manner described.

The program now prompts you to type S to sort another list or Q to quit. You can cycle through as many more examples as you wish by using either random data or data that you type. The program does not repeat the same number in a list when you use the R option, but you can see the algorithm work on a list that does contain several occurrences of the same number by using the T option and typing the list yourself.

The algorithm proceeds in the so-called divide-and-conquer fashion. If a list consists of a single element, the program infers that it is in order. If a list contains more than one element, the rightmost element is used as pivot. If all the other elements are less than the pivot, then the pivot is in place. If all the other elements are larger, the first of these is swapped with the pivot, thereby moving the pivot into place. In all other cases, the leftmost element that is larger and the rightmost element that is smaller are found and swapped. If these were not adjacent, the process described in the preceding sentence is applied to the elements between them. Otherwise, the pivot is swapped with the larger of the two adjacent elements that just were swapped.

When the pivot is in place, action continues as follows in the usual form of the algorithm. The sublist on the left of the pivot (if there is one) is sorted, and then the sublist on the right of the pivot (if there is one) is sorted. Each sublist is sorted by the process just described for the entire list. Stated in this way, the process is recursive. The QWIKSORT program, however, proceeds as follows. When the pivot is in place, it finds the leftmost sequence of elements of the original list that are not yet known to be in place. Then it treats them as the next sublist to be sorted by the overall process which has been described. This makes the process iterative, and the program QWIKSORT keeps track of the iterations in a very simple manner.

The process can be speeded by checking the list at the outset to see if it is in order and doing likewise whenever a sublist is tackled. Other refinements have been explored extensively. Sources of further information are given in the bibliography.

Program Variables

L	The number of items to be sorted
R$	The response to certain prompts
X(i)	The item currently in the ith position of the list
C(i)	The foreground color used to display X(i)
DONE$	A string of length L in which the ith character (a) is 0 initially and (b) is changed to 1 when the ith position in the list is occupied by the number that belongs there in the final arrangement
VCP	The vertical cursor position of the current line of the display
LB,RB	The horizontal cursor positions of the left and right brackets that mark the sublist under current attention
I	The index of successive items in the X array
J,K	The indexes of the leftmost and rightmost items, respectively, of the sublist under current attention (K, therefore, is the index of the pivot)
L,M	The index of an item compared with the pivot in a left-to-right and a right-to-left sweep, respectively
ARL$	The string that consists of the bottom line of arrow markers on the screen suitably spaced for redisplay at the top of the screen after it has been cleared and the display restarted
II	The horizontal position of a character in the line of arrow markers

Program Notes

The string DONE$ that controls the iteration is initialized by

```
310 DONE$=STRING$(L, "0")
```

The algorithm is embodied in a WHILE . . . WEND loop. When the element in position J is in its final place, the program sets the Jth byte of DONE$ to 1 and reaches the WEND statement.

```
1010  WHILE INSTR(DONE$, "0")
 . . .
1840      MID$(DONE$,M,1)="1"
 . . .
1940  WEND
```

When the program has established that all the elements are in place, DONE$ consists entirely of 1's and the WHILE sends control to the statement following the WEND. This conjoint usage of WHILE . . . WEND, INSTR, and the MID$ pseudovariable is quite convenient for a variety of iterative "casting out" and "filling in" processes.

An inner WHILE . . . WEND loop makes the successive comparisons and transpositions proceed until the time is ripe to move a pivot (or to leave it unmoved).

```
1240  WHILE M<=N+1
 . . .
1720  WEND
```

The program jumps past the end of this loop when the current pivot has been moved into place or has been found to be in place already. A third-level WHILE . . . WEND loop controls the left-to-right sweep.

```
1310      WHILE X(M)<=X(K) AND M<=N
 . . .
1370          M=M+1
1380      WEND
```

Another, similar loop controls the right-to-left sweep.

Suggested Activities

1. Check that QWIKSORT works on lists in which not all the items are different.
2. Adapt the QWIKSORT program to print histories of its action on

different sets of data in the style used for the example of a sort of 25 elements. (*Hint:* Use a string S$ that consists of the items in their current order with the brackets [] in place. Use another string DUN$ that contains underscores at the positions of items that are known to be in place. Initialize DUN$ and update it by the statements

```
315    DUN$=SPACE$(3*L+1)
1845   MID$(DUN$,3*M-1,2)="--"
```

Construct S$, and print S$ and DUN$ by

```
1062 S$=" "
1063 FOR I=1 TO L
1064   T$=STR$(X(I)): S$=S$+SPACE$(3-LEN(T$))+T$
1065 NEXT I
1066 S$=S$+" ": MID$(S$,LB,1)="[": MID$(S$,RB,1)="]"
1067 LPRINT S$: LPRINT CHR$(27)+CHR$(10)+DUN$
1068 LPRINT CHR$(27)+CHR$(10)+SPACE$(3*K-LEN(STR$(X(K))))+STR$(X(K))
1069 LPRINT
```

Repeat the instructions to form and to print S$ between lines 1940 and 1950 also. The ASCII codes 27 and 10 at the start of a print line provide a carriage return without a line feed. Consequently, the LPRINT instruction that prints DUN$ superimposes the print position of each underscore with the character that it underlines, giving the correct visual effect. The instruction 1068 overstrikes the pivotal element, correspondingly, to give the effect of boldface type.)

3. Adapt QWIKSORT to print the detailed history of a sorting operation, containing all the arrow markers, that appears on the screen.

4. Write a pointer sort program for character strings that is based on the quicksort algorithm.

5. Apply the dancing numbers technique of BUBLPLUS in a program to demonstrate the quicksort algorithm.

1.5   Some General Considerations

The importance of sorting algorithms is shown by the estimate that sorting occupies about 25 percent of all computer time [Goodman and Hedetniemi, 1977, p. 207]. The speed and cost of computer applications that include sorting operations can be critically affected by the choice of algorithm and the manner in which it is programmed. In some situations you can simply take a standard sort utility for granted. If necessary, you can use further programs to operate mechanically on the data before and after it is sorted.

You are most likely to program a sorting operation when it is tightly integrated in a larger process or when the working environment presents special needs or opportunities. These include particular characteristics of the equipment, the language in which the application is programmed, the data, and the overall systems design. Occasionally, you may need to write a sort program to fill a gap in the system software for a new microcomputer.

In selecting an algorithm, the primary decision depends on the trade-off between simplicity and efficiency. If the individual files of data are small, the insertion sort is likely to be a reasonable choice. It is simple and concise, easy to code and to test, and somewhat better than the selection and bubble sorts for many starting arrangements of the data. All of these elementary methods are better for lists that are already in order than the more advanced algorithms such as quicksort. The weakness of the elementary algorithms of Section 1.1 is that, on average, the time to sort N records increases in proportion to the square of N when N is large. The performance is improved considerably in the *shellsort* algorithm, which is outlined later in this section. Shellsort also is easy to program and to validate, and several authors recommend it in medium-size situations for which the completely elementary algorithms are too slow.

Quicksort has an average processing time that is proportional to N log N. This is even faster than shellsort. Unfortunately, certain initial arrangements of data make the running time of quicksort proportional to the square of N. When average performance is the main concern, you can get the best of both worlds by tempering the use of quicksort with the subsidiary use of insertion sort on sublists that are less than 10 elements long.

Another algorithm, called *heapsort,* also has an average time proportional to N log N, but the proportionality factor is about twice that of quicksort. The processing time varies only slightly around the average, however. Its use may be preferable when catastrophe will strike if the sort is not completed within a restricted time window, but no major benefits result from finishing the sort more rapidly.

Up to now we have assumed that the sorting is *internal,* that is, that all the records are in the main memory of the computer at the same time. Otherwise, the sorting is *external.* Often, an external sort is performed by (1) splitting the complete file into pieces that are put in order separately by internal sorting and saved on disk and (2) merging the results. A merging program is described in Section 2.3. Another external sorting technique that involves merging is outlined in Section 2.4.

Details of the way in which data is gathered can negate the benefits of an advanced algorithm even when the files are long. Thus, if the data is in fairly good order to begin with, an elementary algorithm can deal with it quite rapidly. For an extensive and ongoing application, an analysis of typical data may reveal elements of nonrandomness that can be exploited. One weakness of using theoretical analyses of average behavior is that averages are often rather poor statistical measures. Introductory books on statistics discuss this issue quite extensively and explain other measures such as the median and percentiles.

The theoretical analyses are based on estimating how many comparisons, transpositions, and other moves of data are made on average and in the most unfavorable circumstances possible. Experiments can be run on random data to test if the average time for files of a particular length settles down as the number of trials increases. By conducting these experiments for files of systematically varying length, a curve can be plotted and compared with theoretical expectations.

There is considerable scope for further studies that begin by defining a measure of disorder in a list, such as the minimum number of transpositions needed to put the list in order. Experiments then could be run to time or count elementary operations in the sorting of lists that initially are disordered to varying extents. The distance between the elements that must be transposed and the lengths of sublists that are in order or in reverse order also bear on possible measures of disorder. The quantitative measurement of disorder is by no means a new idea. It has been of central importance in the physics of metals for a considerable time.

It also is of interest to investigate the efficiency of the algorithms that are based on operations to move, reverse, separate, and merge entire sublists that are already in order. The cost of a block operation could be assessed, for example, by using a constant term plus another term proportional to the block size. This approach ties in with some topics in artificial intelligence and with the simulation of actual physical systems in which objects are rearranged. Train shunting is one example. Another example is rearranging a line of objects that are under cover when at rest but under fire when being moved. If the chances of being hit go up rapidly with the distance of each move, but are finite no matter how short the move, some interesting trade-off strategies result.

In an ongoing application that builds files of increasing size, the overall system design can create or avoid increasingly massive sorting operations. Thus, it is wasteful to update a lengthy sorted file by appending a few new records and sorting the resulting file completely. A far better approach is to sort the new batch of records and merge the result into the original file. Suppose too that a file is updated periodically, processed in alphabetic order of a field that contains a personal name, and then processed in numeric order of a field that contains a social security number. If external storage is inexpensive and time is at a premium, it may be advantageous to maintain one version of the file sorted by name and another version sorted by number. Then each of these versions can be updated by merging with the suitably sorted batch of new records. When this approach is taken, however, cast-iron working procedures must be established to ensure that if a record is corrected in one file, then the same correction is made and validated in the other file, too.

Still further considerations affect the expense and the elapsed processing time when a large time-shared computer is used to run a sorting utility or an application program that contains a sorting component. The relative charges for the use of main storage and for input/output operations determine the optimal size of the pieces of a very large file that are sorted internally, stored on disk, and then merged. These factors may warrant the use of a very large region, but if waiting for this to become available can cause serious delays, a more modest approach is preferred when turnaround time is a major concern.

Thus the optimal way to handle a substantial sorting application requires considerable thought, analysis, and downright experimentation. The choice of algorithm is a critical element, but it can be swamped by other considerations including overall systems design.

A very simple algorithm that must not be overlooked applies when the only values that are possible for the keys belong to a small set of alternatives and an individual list cannot contain the same key more than once. In that case, the list of all possible key values, arranged in order, can be mapped onto a set of consecutive integers 1 to N. An array of records can then be sorted simply by putting the addresses in an array that contains one slot for each possible key value.

The shellsort algorithm mentioned earlier starts by correcting the order of the items that are relatively far apart and then corrects the disorder among the items that are closer together. Thus, if the list contains 100 items, the first step uses the insertion sort to put the items in positions 1, 14, 27, 40, 53, 66, 79, and 92 in order relative to each other, without reference to the rest of the list. These items are called the subsequence with interval 13 starting at 1, because the first item is in position 1 and the positions of the others increase by 13. Then the items in positions 2, 15, . . . , 93 are put in order. These constitute the subsequence of interval 13 starting at position 2.

The remaining subsequences of interval 13, starting at positions 3,

4, . . . , 13, are put in order separately. These steps impose a strong degree of order on the list. To refine it, the items in positions 1, 5, 9, . . . are put in order. These constitute the subsequence of interval 4, starting at 1. The subsequences of interval 4, starting at positions 2, 3, and 4 are put in order. Finally, the insertion sort is applied to the resulting list to make any residual changes that are needed. Usually, there are very few, and the entire process is much faster than applying the insertion sort to the original list.

In this example, the subsequence intervals 13, 4, and 1 were used in turn. These are taken in reverse order from the series 1, 4, 13, 40, 121, 364, ... Each of these numbers is 3 times its predecessor plus 1. In general, the largest number in this series that is less than or equal to the length of the list is used for the first subsequence interval, then the preceding number in the series, and so on to 1. Other series have been used, but the one just cited has been found to be quite effective. In principle, the subsequences can be sorted by any algorithm, but the insertion sort usually is employed.

The several books cited in the bibliography provide detailed analyses and further discussions of the algorithms that we have described. These books also describe several other algorithms for sorting and some related processes. They include radix sorting, which conceptually resembles the principle of the mechanical punched-card sorter invented by Herman Hollerith nearly 100 years ago. A similar principle is used to sort a file by using a hierarchy of keys. Suppose, for example, that a file deals with animals in a zoo. Each record contains the family, genus, and species of an animal, together with other relevant data. Suppose that we want the file sorted alphabetically by species within genus within family. In other words, all the records for members of the same genus are brought together, and all the groups of records for genera within the same family are brought together. The family groups are arranged in alphabetic order by family name. Within each family, the genus groups proceed in alphabetic order by genus name. Within each genus, the species names proceed in alphabetic order also.

The zoo file can be sorted into this order with any of the algorithms we have described by using the concatenation of the family name, genus name, and species name as the key. Alternatively, the comparison of two records can proceed piecewise and transposition performed when necessary. First, the family names are compared and the records (or pointers) are left unchanged or transposed if the names are in order or out of order, respectively. If the family names are the same, however, the genus names are compared. A transposition is made if they are out of order or if they are the same and the species names are out of order.

When there are a lot of species in the same genus and a lot of genera in the same family, both the processes that were just described compare the family names and genus names far more often than is necessary. It is much more efficient, in this circumstance, to sort the file on the species key, then on the genus key, and then on the family key by using the insertion sort or shellsort. These methods are stable in the sense that they preserve the order of records which have the same value of a key. Bubble sort also is stable, but the selection sort, quicksort, and most of the other advanced methods are not. Radix sorting uses successive digits from right to left in a numeric key. Its benefit is related to the circumstances we have just discussed.

Suggested Activities

1. Elaborate the bubble sort algorithm as follows:
   (a). Adopt a look-ahead tactic before transposing. That is, when sorting into ascending order, if $x(m) > x(m+1)$, $x(m+1) \leq x(m+2) \leq \cdots \leq x(m+k)$, and $x(m+k) < x(m)$, then transpose $x(m)$ and $x(m+k)$. This avoids unnecessary transpositions in a portion of the list that is in order and which is sandwiched between two elements that should be transposed around it, for example 7, 2, 3, 4, 5, 6, 1.
   (b). Adopt a look-back tactic such that, whenever the items in two successive slots are found to be out of order, the longest possible block that ends with the first item is moved around the second item. For example, the first such block move in the list 3, 6, 7, 4, 5 rearranges it to 3, 4, 6, 7, 5.
   (c). Superimpose the look-ahead tactic on the look-back tactic, so that, for example, the first block move in the list 3, 7, 8, 10, 4, 5, 9 rearranges it to 3, 4, 5, 7, 8, 10, 9.
   Relate these elaborations to other algorithms that have been discussed.

2. A list of n items can be reversed by int (n/2) transpositions. Explore the potential for sorting algorithms that (a) search the list being sorted for lengthy sublists that are in reverse order, (b) reverse the sublists, and then (c) apply the look-ahead and look-back tactics described in Activity 1.

3. Write a program to demonstrate shellsort.

4. (a). Expand the SIMPSORT program to count the number of comparisons and transpositions and other moves that are made on different samples of data when each of the three elementary algorithms is used.
   (b). Expand Activity 2 in a corresponding manner.

5. Repeat Activity 4(a) for the QWIKSORT program.

6. Repeat Activity 4(a) for a shellsort program.

7. Write a program to simulate a train shunting or a "sort under fire" situation, such as we described, to experiment with optimal strategies.

8. Review the mathematical analysis and the associated discussion of sorting algorithms. Several major sources of information are cited in the bibliography.

2

Looking for a Match

In this chapter we discuss some ways of using files that have been sorted. Sections 2.1 and 2.2 deal with methods for searching files of data that can be used, for example, to retrieve information and to process business transactions. Section 2.3 shows how pairs of files can be compared, item by item, to produce a single merged file or to remove items from one file that occur in the other.

Keys are used in searching, merging, and related operations just as they are in sorting. Thus a computer search typically yields a record (or several records) in which the content of a particular field matches the key that was specified. Merging combines files that have been sorted on the same key to produce a new file that is sequenced on that key also.

You may want to retrieve the record that contains a particular value in the key field because you need the other data in the record for use outside the computer. Thus, you may feed the name of a friend into a computer that searches a file of personal information to get his or her telephone number. Or you may want to retrieve a record so that you can alter it and put the revised form back in the file. That operation is carried out, for example, to update an inventory or credit balance of a customer after a sale is made or to update the record that contains the address and phone number of a friend after he or she has moved. Files are merged for many reasons; an example is to consolidate the phone books of the divisions of a corporation into a single company-wide book.

2.1 Performing a Binary Search

When you use a computer to retrieve information such as the telephone number of a friend, you type your friend's name and the computer locates the name in a file of reference data. The program that locates the name may do so by checking each successive name in the file until either (1) it finds a match for the name that you typed or (2) it reaches the end of the file. This is called a sequential search. The item that you typed is called the *search key*. If the file contains a million items and the search key matches the last of them, the computer performs a million comparisons before it finds what you are looking for. It performs a million comparisons also whenever the search key does not match any of the items in the file.

The number of comparisons can be reduced when the file has been so arranged that the values with which the search key is compared are in increasing numerical order or are in alphabetic order. If the search key is absent from the file, the program can detect its absence on reaching an item that is larger if the key is numeric or which follows the key alphabetically if the key is a string.

The process can be made even more efficient by using the binary search technique that the program BINHUNT demonstrates. BINHUNT lets you search a list of random numbers that it generates in numerically increasing sequence as a preliminary. The program begins by prompting for the length of the list to be searched. Start by requesting 20 numbers. Later on you may want to experiment with longer lists, but you can learn a lot from the program by starting small. The program sets a limit of 999 on the length of the list, but this can be increased very easily. After you respond, the program generates a list of integers that starts with a random value between 1 and 5 and increases from item to item by a random amount between 1 and 50. When the list has been constructed, the program prompts you to type D to display the list or S to start searching. Type D to display the list. If you requested 20 numbers, the following two-column display appears on the right side of the screen:

1	4
2	39
3	64
4	114
5	147
6	154
7	173
8	203
9	210
10	257
11	268
12	284
13	296
14	341
15	371
16	395
17	396
18	438
19	440
20	459

Thus, the items appear together with the "slot numbers" (positions in the list). Then a prompt for the search key appears at the top left of the screen. Respond by entering the value 341. The top line of the display changes to:

```
Search key: 341 - press space bar to start searching
```

You press the space bar, and the top-left portion of the display changes to:

```
Search key: 341
```

test no.	items under attention	position and value of item to be compared	search key is
1			

Press space bar to show range under attention

You press the space bar again. The program fills in the next blank position of the table to show that at this point the entire list, that is, the first to twentieth items, has to be considered.

```
Search key: 341
```

test no.	items under attention	position and value of item to be compared	search key is
1	1st- 20th		

Press space bar to select middle item

You press the space bar, and the program picks the middle item, that is, the tenth. It has the value 257. The program fills in the next blank position in the table accordingly.

```
Search key: 341
```

test no.	items under attention	position and value of item to be compared	search key is
1	1st- 20th	10th 257	

Press space bar to compare

The program compares the key, 341, with the tenth item and finds that the key is greater. The final entry in the first line of the table is filled in.

Search key: 341

test no.	items under attention	position and value of item to be compared		search key is
1	1st- 20th	10th	257	greater

Press space bar to start next cycle

By finding that the key exceeds the tenth item, the program has eliminated half of the list from further consideration. You press the space bar again and the table expands to show the progress of the next cycle.

Search key: 341

test no.	items under attention	position and value of item to be compared		search key is
1	1st- 20th	10th	257	greater
2				

Press space bar to show range under attention

You press the space bar, and the program shows that the new range consists of just the eleventh to twentieth items.

Search key: 341

test no.	items under attention	position and value of item to be compared		search key is
1	1st- 20th	10th	257	greater
2	11th- 20th			

Press space bar to select middle item

You respond, and the program shows that the key will be compared with the fifteenth item, which is 371.

Search key: 341

test no.	items under attention	position and value of item to be compared		search key is
1	1st- 20th	10th	257	greater
2	11th- 20th	15th	371	

Press space bar to compare

You press the space bar to compare, and the program shows the key is
smaller.

Search key: 341

test no.	items under attention	position and value of item to be compared		search key is
1	1st- 20th	10th	257	greater
2	11th- 20th	15th	371	smaller

Press space bar to start next cycle

The search has been narrowed to just the eleventh to fourteenth items.
The program prompts you to press the space bar to start the next cycle.
You respond, and the table expands accordingly. The program repeats
the sequence of prompts that it gave in each of the first two cycles. This
leads to the table containing the following information.

Search key: 341

test no.	items under attention	position and value of item to be compared		search key is
1	1st- 20th	10th	257	greater
2	11th- 20th	15th	371	smaller
3	11th- 14th	12th	284	greater

Press space bar to start next cycle

Now the search is narrowed to just the thirteenth and fourteenth
items. The program repeats the sequence of prompts which it gave in
earlier cycles. The table expands to the following:

Search key: 341

test no.	items under attention	position and value of item to be compared		search key is
1	1st- 20th	10th	257	greater
2	11th- 20th	15th	371	smaller
3	11th- 14th	12th	284	greater
4	13th- 14th	13th	296	greater

Press space bar to start next cycle

All that is left now is to try the fourteenth item. The program does this step by step, in response to successive key presses, and finds that the key matches. The left part of the display is now:

Search key: 341

test no.	items under attention	position and value of item to be compared		search key is
1	1st- 20th	10th	257	greater
2	11th- 20th	15th	371	smaller
3	11th- 14th	12th	284	greater
4	13th- 14th	13th	296	greater
5	14th	14th	341	equal

The search key matches the 14th item

Thus, the program found the key after 5 comparisons instead of the 14 comparisons needed in a sequential search. With 20 items in the list, the greatest number of comparisons needed to find the key, or to show it is absent, is 5.

The program prompts you to type A for another search, N for a new list, or Q to quit. To explore the operation of the algorithm some more, type A for another search. The screen clears, and you are again prompted for a search key. This time type 3. The search proceeds through a succession of steps that are recorded in the table which is on the screen at the end.

Search key: 3

test no.	items under attention	position and value of item to be compared		search key is
1	1st- 20th	10th	257	smaller
2	1st- 9th	5th	147	smaller
3	1st- 4th	2nd	39	smaller
4	1st	1st	4	smaller

The search key is less than the first item

Thus the algorithm narrowed the range to items 1 to 9, then to items 1 to 4, and then to the first item in successive cycles. Correspondingly, if you make the search key 500, the table at the end of the process shows that the range was narrowed in turn to items 11 to 20, then to items 16 to 20, then to items 19 to 20, and finally to item 20.

Search key: 500

test no.	items under attention	position and value of item to be compared		search key is
1	1st- 20th	10th	257	greater
2	11th- 20th	15th	371	greater
3	16th- 20th	18th	438	greater
4	19th- 20th	19th	440	greater
5	20th	20th	459	greater

The search key is more than the final item

The last two searches took four and five comparisons, respectively. Quite often, however, still fewer comparisons are needed. Thus if the search key is 147, the algorithm just takes two cycles to find that it matches the fifth item in the list.

Search key: 147

test no.	items under attention	position and value of item to be compared		search key is
1	1st- 20th	10th	257	smaller
2	1st- 9th	5th	147	equal

The search key matches the 5th item

Searching for 70 illustrates the behavior when the key falls between two items in the list. The display at the end of the process is:

Search key: 70

test no.	items under attention	position and value of item to be compared		search key is
1	1st- 20th	10th	257	smaller
2	1st- 9th	5th	147	smaller
3	1st- 4th	2nd	39	greater
4	3rd- 4th	3rd	64	greater
5	4th	4th	114	smaller

The search key is between the 3rd and 4th items

The general principle of the binary search thus begins by starting in the middle of the list. The item you are seeking meets one of three conditions:

1. It matches the middle item of the file.

2. It precedes the middle item alphabetically or numerically.

3. It comes after the middle item.

In case 1 the search is ended. In case 2 you can ignore everything in the second half of the file immediately. In case 3 you can ignore everything in the first half. One comparison thus finds the required item if you are lucky or, at worst, halves the amount of searching that remains. To continue, you compare the key with the item in the middle of the portion of the file to which the search was narrowed by the first comparison. Either you score a hit in this cycle or you narrow things down to a quarter of the original file. Continuing in this way, successive comparisons confine attention to an eighth, a sixteenth, and a thirty-second of the original file, after three, four, and five comparisons, respectively, if the search key is not found in the process.

The worst case that can occur with a file of a million entries is that you take 20 comparisons to locate what you are looking for or to find that it is not in the file. In fact, 20 comparisons are enough for a file of 1,048,575 entries, or 2 to the power 20 minus 1. In general, n comparisons are the most that are needed to locate an item in a file of 2 to the n minus 1 entries (or to show that the item is missing). Because each comparison, up to the time that the search ends, divides the portion remaining under scrutiny by 2, this process is called a binary search.

You can experiment quite extensively with the BINHUNT program. To change the list, you type N for a new list in response to the prompt that follows a search. When a list of more than 24 items is displayed on the screen by typing D in response to the appropriate prompt, the screen scrolls. You can freeze the display at a point that you want to keep under attention by pressing the space bar. Typing Q in response to the prompt that follows a search makes the program end. The storage capacity of the computer limits the length of the list that the program generates, but the time needed to generate a list of more than a few hundred entries becomes noticeable before storage becomes the limiting factor.

Program Variables

HCL	The horizontal cursor position of the slot number of an item
B$	The graphics character |
H$	The string of hyphens used in the table on the screen
RW$(i)	The word "smaller," "equal," or "greater" when i is 1, 2, and 3, respectively

M	The position of the item in the list that is compared with the search key in the current cycle
IMAX	The length of the list that is generated and searched
N%(i)	The ith item in the list
I	The subscript of an item in the list
R$	The response to certain prompts
JLO	The lower limit of the portion of the list to which the search is currently confined
JHI	The upper limit of the above portion of the list
D	The index of a delay loop
FN D(M)	The unit digit of M unless M is in the range 10 to 20, in which case the value of the function is 0
FN T(M)	The number 1, 3, 5, and 7 when the suffix of the ordinal form of M is st (as in 1st), nd (as in 2nd), rd (as in 3rd), and th (as in 4th), respectively
FN TAG$(M)	The actual two-letter suffix st, nd, rd, or th
FN DS$(M)	The string representation of M without a leading space
FN ORD$(M)	The ordinal adjective corresponding to M, for example, 1st or 23rd
NQ	The search key
K	The cycle number in the iteration
REL	The switch that is 1, 2, or 3 when the key is less than, equal to, or greater than the item with which it was compared most recently

Program Notes

The search is performed by a loop that cycles until one or the other of these events:

1. The rule to form JHI and JLO for the next cycle makes JHI<JLO, showing that the requested value is not in the list.

2. The switch REL is set to 2 because the key matched the item with which it was just compared.

Accordingly, the algorithm is embodied in the control structure by the following statements:

```
1010   WHILE JLO<=JHI AND REL<>2
1530     REL=SGN(NQ-N%(M))+2
1560    IF REL=1 THEN JHI=M-1 ELSE IF REL=3 THEN JLO=M+1
1570   WEND
```

```
100 REM * BINHUNT binary search demonstration *
101 :
110  ON ERROR GOTO 5000                                          'error trap
120  KEY OFF: SCREEN 0: WIDTH 80                            'set up the screen
130  COLOR 7,0: CLS
140  HCL=61                                              'tab position of list
150  B$=CHR$(179)                                               'vertical bar
160  H$=STRING$(52,"-")                                       'line of hyphens
170  RW$(1)="smaller": RW$(2)="equal": RW$(3)="greater"
199 :
200 REM * functions to form ordinals from numbers *
210  DEF FN D(M)=-((M<10)+(20<M))*(M-10*INT(M/10))       'allow for 11th-19th
220  DEF FN T(M)= -1*(FN D(M)=1) -3*(FN D(M)=2) -5*(FN D(M)=3)
                  -7*(FN D(M)=0 OR FN D(M)>3)               'ending selector
230  DEF FN TAG$(M)=MID$("stndrdth",FN T(M),2)               'select the ending
240  DEF FN DS$(M)=MID$(STR$(M),2)
250  DEF FN ORD$(M)=FN DS$(M)+FN TAG$(M)
299 :
300 REM * construct the list to be searched *
310  GOSUB 4000                                            'prompt for length
320  INPUT; "Length of list to be searched: ",IMAX
321  IF 0<IMAX AND IMAX<1000 THEN 330
322  GOSUB 4000
323  PRINT "Please use a length in the range 1 to 999. ";
324  PRINT "Press space bar and try again."
325  Z$=INPUT$(1)
326  GOTO 310
330   DIM N%(IMAX)                                           'allocate space
340   N%(1)=1+INT(5*RND)                                    'starting value
350   FOR I=2 TO IMAX                                 'loop through elements
360    N%(I)=N%(I-1)+1+INT(50*RND)                         'random increment
370   NEXT I
399 :
400 REM * display the list optionally *
410  GOSUB 4000                                                      'prompt
420  PRINT "(D)isplay the list or (S)tart searching: ";
430  R$=INPUT$(1)
440  CLS
450  IF R$="S" OR R$="s" THEN 600                                   'branch
460   IF INKEY$<>"" THEN 460                         'clear keyboard buffer
470   FOR I=1 TO IMAX
480    LOCATE CSRLIN,HCL                               'display the item #
490     PRINT USING "### "; I;
500    LOCATE CSRLIN,HCL+5
510     PRINT USING "#### "; N%(I)                              'and value
520    IF INKEY$<>"" THEN 600             'freeze display if key is pressed
530   NEXT I
599 :
600 REM * prompt and begin a search *
610  LOCATE 1,1                                   'prompt for search key
620   INPUT; "Search key: ",NQ
625    IF NQ=0 THEN LOCATE 1,12: PRINT 0;          'allow for null response
630   JLO=1                                        'end #s of initial range
640   JHI=IMAX
650   HCP=POS(0)                                          'cursor position
660   PRINT " - press space bar to start searching";           'prompt
670   Z$=INPUT$(1)
680   LOCATE 1,HCP                                       'erase the prompt
690    PRINT TAB(HCL-1)
700   LOCATE 3,1                                 'display the column headings
```

```
710    PRINT H$
720    PRINT B$ "test" B$ " items under " B$
                        "position and value of" B$ " search  " B$
730    PRINT B$ "no. " B$   attention   " B$
                        " item to be compared " B$ " key is  " B$
740    PRINT H$
750    K=Ø                                                  'cycle #
760    REL=Ø                                      'comparison switch
999  :
1ØØØ REM * main search cycle *
1Ø1Ø   WHILE JLO<=JHI AND REL<>2      'loop until match or absence established
1Ø2Ø    IF K=Ø THEN 1Ø6Ø
1Ø3Ø     LOCATE K+8,1
1Ø4Ø     PRINT "Press space bar to start next cycle"         'prompt
1Ø5Ø     Z$=INPUT$(1)
1Ø6Ø     K=K+1                                               'cycle #
1Ø7Ø     LOCATE K+6,1
1Ø8Ø     PRINT B$ TAB(6) B$ TAB(2Ø) B$ TAB(42) B$ TAB(52) B$     'draw bars
1Ø9Ø     PRINT H$                                            'hyphens
11ØØ     LOCATE K+6,3
111Ø     PRINT USING "##";K;                        'display cycle #
112Ø     M=INT((JLO+JHI)/2)                'position of comparison item
113Ø     GOSUB 4Ø5Ø
114Ø     PRINT "Press space bar to show range under attention";       'prompt
115Ø     Z$=INPUT$(1)
116Ø     LOCATE K+6,8
117Ø     PRINT USING "###"; JLO;           'display item # of start of range
118Ø     PRINT FN TAG$(JLO);
119Ø     IF JHI=JLO THEN 14ØØ
12ØØ     PRINT USING "-###"; JHI;               'item # of end of range
121Ø     PRINT FN TAG$(JHI);
1399 :
14ØØ REM * compare key with item in middle of current range *
141Ø     GOSUB 4Ø5Ø                                'prompt for comparison
142Ø     PRINT "Press space bar to select middle item";
143Ø     Z$=INPUT$(1)
144Ø     LOCATE K+6,22
145Ø     PRINT USING "###";M;              'display item # of comparison item
146Ø     PRINT FN TAG$(M) TAB(34)
147Ø      FOR D=1 TO 2ØØ: NEXT D                              'delay
148Ø     PRINT USING "######"; N%(M)        'display value of comparison item
149Ø     GOSUB 4Ø5Ø
15ØØ     PRINT "Press space bar to compare";                 'prompt
151Ø     Z$=INPUT$(1)
152Ø     GOSUB 4Ø5Ø                                'clear message area
153Ø     REL=SGN(NQ-N%(M))+2                 'set the comparison switch
154Ø     LOCATE K+6,44
155Ø     PRINT RW$(REL)              'display "smaller", "equal" or "greater"
156Ø     IF REL=1 THEN JHI=M-1 ELSE IF REL=3 THEN JLO=M+1     'change the range
157Ø   WEND
1999 :
2ØØØ REM * key has been matched or shown absent *
2Ø1Ø   PRINT
2Ø2Ø   IF REL=2 THEN PRINT "The search key matches the "
                        FN ORD$(M) " item":        GOTO 2Ø8Ø    'key found
2Ø3Ø   PRINT "The search key is ";
2Ø4Ø   IF REL=1 AND M=1 THEN PRINT "less than the first item":
                                        GOTO 2Ø8Ø       'too low
2Ø5Ø   IF REL=3 AND M=IMAX THEN PRINT "more than the final item":
                                        GOTO 2Ø8Ø       'too high
```

```
2Ø6Ø   IF REL=1 THEN M=M-1                                              'sandwiched
2Ø7Ø   PRINT "between the " FN ORD$(M) " and " FN ORD$(M+1) " items"
2Ø8Ø        Z$=INPUT$(1)
2999 :
3ØØØ REM * prompt for further action *
3Ø1Ø   PRINT
3Ø2Ø    PRINT "(A)nother search, (N)ew list or (Q)uit: ";              'prompt
3Ø3Ø    R$=INPUT$(1)
3Ø4Ø      CLS
3Ø5Ø     IF R$="N" OR R$="n" THEN ERASE N%: GOTO 3ØØ                    'new list
3Ø6Ø      IF R$="Q" OR R$="q" THEN 6ØØØ ELSE 4ØØ                  'quit or search
3999 :
4ØØØ REM * clear the message area *
4Ø1Ø   LOCATE 1,1                                               'clear top line
4Ø2Ø    PRINT TAB(8Ø)
4Ø3Ø   LOCATE 1,1
4Ø4Ø   RETURN
4Ø5Ø REM * clear the prompt line *
4Ø6Ø   LOCATE K+8,1                                       'clear line below table
4Ø7Ø    PRINT TAB(HCL-1)
4Ø8Ø   LOCATE K+8,1
4Ø9Ø   RETURN
4999 :
5ØØØ REM * error trap *
5Ø1Ø   PRINT
5Ø2Ø    PRINT "Error" ERR "in line" ERL                         'error diagnostic
5Ø3Ø     RESUME 6ØØØ
5999 :
6ØØØ REM * quit *
6Ø1Ø   LOCATE 2Ø,1
6Ø2Ø   ON ERROR GOTO Ø                                       'reset error address
6Ø3Ø   END                                                                'quit
```

Suggested Activities

1. Adapt the BINHUNT program (a) to search a list of numbers that is in decreasing numeric order, (b) to print a table that shows the history of a completed search, and (c) to allow the list to be 9999 items long.

2. Adapt the BINHUNT program to search a list of words or phrases that are in alphabetic order. Set up the array that contains the words or phrases, when the program starts to execute, by transferring them from DATA statements or by reading them from a file on disk.

3. When you look for an item in a list, you tend to start by looking near the beginning if the value of the item is close to that of the first element in the list, and you start looking near the end if the value of the item is close to that of the final element. Write a program that is based on this principle. (*Hint:* If you use the variables defined for BINHUNT, set M, the subscript of the element that is compared with the key, to

$$JLO + INT((NQ-N\%(JLO) + 1)/(N\%(JHI)-N\%(JLO) + 1))$$

in each cycle of the WHILE . . . WEND loop. This turns the process into an interpolation search. See how much it cuts down the number of cycles.)

4. You can extend the interpolation search principle to a list of strings in the following way. If the first and final strings in the portion of the list that is under attention begin with different letters, find where the first letter of the key would fit between them assuming that each intervening letter begins the same number of strings. If the first and final elements begin with the same letter, apply this principle by using the second letter, and so on. Use the ASCII codes in the interpolation formula; that is, take ASC(MID$(a$,i,1)) as the number to use for the ith character of the string a$ when applying the formula.

5. Write a program to search a series of ranges for a key. For example, if the ranges are 3 to 7, 15 to 18, 28 to 34, . . . , and the key is 29, the program would show that key is in the third range. If the key were 22, the program would show that the key fell between the second and third ranges.

6. Suppose you have a long list of items that are almost in order. You can reduce the number of comparisons in an insertion sort by using a binary search to find where to move the item under current attention. Write a program that adopts this tactic. Modify it to use the interpolation search.

2.2 Zooming In through Several Levels

The program MULDIREC illustrates another technique that is used in information storage and retrieval. It employs a two-level directory to display the names of countries in different continents and the names of towns in different countries. The information is displayed through two levels of window. The program provides a prototype of the browsing feature that is found in many information retrieval systems. Here you will find a small body of information that you can easily expand to increase the number of continents, countries, and towns that are listed. You can also alter the program to retrieve voluminous data from disk and display it.

The program begins by displaying an explanation of how it is used. Then it draws the schematic of a small geographic data base (Figure 2.1). The data base consists of (1) a list of towns grouped hierarchically by country and continent, (2) a list of countries grouped by continent, and (3) a list of continents. These are depicted by three boxes in the schematic. The box for the continent list can be read. The other boxes

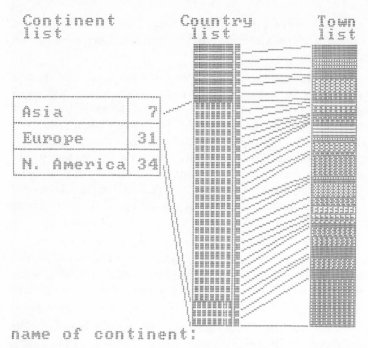

Figure 2.1

represent information that is too extensive to be displayed in a readable manner all at once.

The data in the continent list shows that the ends of the sections for Asia, Europe, and North America in the country list occupy slots 7, 31, and 34, respectively. Because the first section starts in slot 1, and each later section starts in the slot that follows the end of the preceding section, the sections for the three continents occupy slots 1 to 7, 8 to 31, and 32 to 34, respectively.

After displaying the schematic, the program prompts for the name of a continent. Suppose you type Asia. The Asian section of the country list is displayed in regular-size characters in a window that cuts through part of the continent list and through part of the stylized country list. The data in the window can also be likened to a placard placed in front of the other information (Figure 2.2). The first line of information in the window shows that the final entry for Burma is item 1 of the town list; that is, only 1 town in Burma is included. The final entry for China is item 7 of the town list. Thus, items 2 to 7 are the names of towns in China. Items 8 to 12 of the town list are names of towns in India, and so on.

Simultaneously with the "explosion" of the Asian section of the

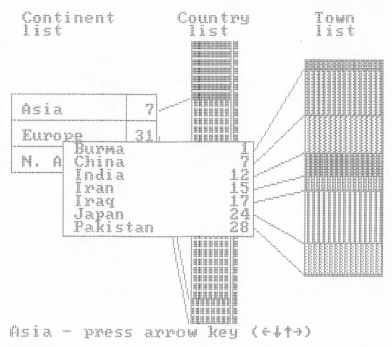

Figure 2.2

country list, a box that depicts an explosion of the corresponding part of the town list is displayed on the right-hand side of the screen, cutting through (or placed in front of) the information already there. This is divided into sections for the different countries in Asia. The heights of successive sections are in proportion to the numbers of towns in the respective countries for which data has been provided to the program.

You now have a choice of four actions that are brought into play by the four arrows on the numeric keypad on the right-hand side of the keyboard. Pressing the left arrow revokes the most recent change and thereby restores the display shown in Figure 2.1. Pressing the down arrow replaces the Asian section of the country list by the European section. Pressing the up arrow replaces the Asian section by the section for North America. Thus the down and up arrows move down and up, respectively, in the continent list and select and display the appropriate section of the country list. The arrows produce a wraparound effect from the bottom of a list to the top, and vice versa.

Pressing the right arrow moves to the next level of detail. The prompt at the bottom of the screen changes to:

```
name of country:
```

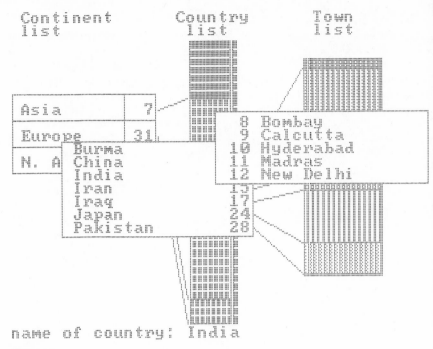

Figure 2.3

Suppose you enter India in response. The names of the towns in India, which were provided to the program, appear in yet another window cutting through the material on the screen. They are Bombay, Calcutta, Hyderabad, Madras, and New Delhi. They occupy slots 8 to 12 in the entire list of towns. This is consistent with the data in the Asian section of the country list, which is still partly visible (Figure 2.3).

If you would like to see the names of towns in other Asian countries, you can press the down arrow to work downward through the Asian section of the country list. Thus, you see the towns in Iran, Iraq, and so on. Pressing the up arrow works upward from India to China to Burma. Pressing the left arrow restores the display shown in Figure 2.2. You can proceed to display the sections of the country list that belong to the different continents in the manner already described. Thus, in general you press the right and left arrows to increase or decrease the level of detail that is displayed, that is, to zoom in and out. You press the down and up arrows to review more data at the current level of detail, that is, to pan. You can quit by pressing Q after a list of towns has been displayed.

The program displays successive groups of items in a window when the entire list of countries in a continent or of towns in a country would not fit. You press the space bar to proceed from each group to the next. Pressing the Enter key when prompted for a continent or a country has the same effect as entering the name of the first continent or of the first country in the continent under attention. The display that appears at the start of execution, to provide a summary explanation of the way the program is used, is as follows:

```
The program first prompts for the name
of a continent. After you respond, it
prompts for an arrow key (→↓↑←).

  You press
  ←  to recall the continent list,
  ↓  for the next continent,
  ↑  for the previous continent,
  →  to specify a country.

After you specify a country, the program
prompts for an arrow key (←↓↑) or Q.

  You press
  ←  to recall the country list,
  ↓  for the next country,
  ↑  for the previous country,
  Q  to quit.

Please wait while data are loaded

Press the space bar to start.
```

Because of their roles in locating entries, the country list serves as a directory of the town list and the continent list serves as a directory of the country list.

Program Variables

XL	The horizontal screen coordinate of the left edge of the schematic lower-level directory
XR	The horizontal screen coordinate of the left edge of the schematic town list in Figure 2.2
U%	The array used to preserve the bit map of the part of the display that is changed when the lower-level directory for the countries in a continent is shown on the screen
V%	The array used to preserve the bit map of the part of the display that is changed when the towns in a specified country are shown on the screen
VC	The vertical cursor position of the count that is displayed when data is loaded at the start of execution
HC	The horizontal cursor position of the count

CONTINENT$(i)	The ith name of a continent in the data provided to the program
COUNTRY$(j)	The jth name of a country in the data
TOWN$(k)	The kth name of a town in the data
I	The index of an item in the array CONTINENT$
J	The index of an item in the array COUNTRY$
K	The index of an item in the array TOWN$
IMAX, JMAX, KMAX	The maximum values of I, J, K for the data provided to the program
P(i)	The pointer to the entry in the COUNTRY$ array for the final country on the ith continent
Q(j)	The pointer to the entry in the TOWN$ array for the final town in the jth country
YL(i)	The vertical screen coordinate of the bottom of the section for the ith continent in the schematic country list
YR1(j)	The vertical screen coordinate of the bottom of the section for the jth country in the schematic country list
YR2(j)	The vertical screen coordinate of the bottom of the section for the jth country in the schematic town list
A$	The item that was read from a DATA statement in the current cycle of the loop that sets up the arrays
L$	The first character of A$
B$	The name that is typed in response to the prompt for a continent
JLO	The J value of the first country to be displayed in the country window
JHI	The J value of the final country to be displayed in the country window
C$	The name that is typed in response to the prompt for a country
YCEXP	The vertical screen coordinate of the section for a particular country in an "exploded" schematic town list—this variable is used in turn for all the countries of the selected continent in a FOR . . . NEXT loop
FNT$	A random two-byte string that defines a tiling pattern
WH	The number of text lines in the window that is being formed
KLO	The K value of the first town to be displayed in the town window
KHI	The K value of the final town to be displayed in the town window

PSW	The switch that is 0 when a lower-level directory or a list of towns is about to be displayed and is 1 after the first group has been displayed and the directory or the list is not exhausted
YCP	The vertical pixel position of the top of the section of the schematic town list to be displayed next in the overall action that displays the directory of countries in a continent
MA$(t)	The phrases that explain the role of the arrow keys when a continent can be selected
MB$(t)	The corresponding phrases that explain the role of the arrow keys when a country can be selected
W$	The invalid response to a prompt that is displayed in a diagnostic message
AR$,AD$	The one-byte codes for the right arrow and down arrow
AU$,AL$	The codes for the up arrow and left arrow

Program Notes

The names of continents, countries, and towns are provided on a batch of DATA statements. The statements are placed at the end of MULDIREC to permit the inclusion of more extensive data without altering the line numbers of the rest of the program. The characters $ and # precede the names of continents and countries, respectively. The names of all towns in the same country are listed in alphabetic order after the name of the country. The data for all the countries in the same continent is listed after the name of the continent, and the names of the countries are in alphabetic order. The data is so arranged that the names of the continents are in alphabetic order too. The general pattern thus is $Asia, #Burma, Rangoon, #China, Canton, . . . , $Europe, ... To facilitate later expansion of the information, a separate DATA statement is used for each continent name and to start each country

```
15010 DATA $Asia
15020 DATA #Burma, Rangoon
15030 DATA #China, Canton, Changsha, Foochow, Nanking, Peking, Shanghai
```

This data is transferred to the CONTINENT$, COUNTRY$, and TOWN$ arrays during the initialization.

When the name of a continent is typed in response to the prompt, a simple FOR . . . NEXT loop looks for it in the CONTINENT$ array. In the expressions that change the subscript which identifies the continent, in response to an arrow key press, the MOD function is used to wrap around from IMAX to 1 and vice versa.

```
2270 I=(I MOD IMAX)+1: GOTO 2290
2280 I=((I+IMAX-2) MOD IMAX)+1
```

The countries in the Ith continent are stored in slots $P(I-1)+1$ through $P(I)$ of the array COUNTRY\$. Thus, the J value of the country that you specify in response to the prompt is found by the loop

```
3100 FOR J=P(I-1)+1 TO P(I)
3110   IF C$=COUNTRY$(J) THEN 3210
3130 NEXT J
```

Correspondingly, the number of text lines in the window that displays the portion of the lower-level directory pertaining to the continent I is set tentatively by:

```
7040 WH=P(I)-P(I-1)
```

The window height is reset to 13 if the tentative value is greater. The appropriate portion of the lower-level directory then is displayed in groups of 12 by the loop:

```
7200 FOR JLO=P(I-1)+1 TO P(I) STEP 13
. . .
7600 NEXT JLO
```

When the name of a country is typed in response to the prompt, it is sought in the COUNTRY\$ array by the loop that begins:

```
3100 FOR J=P(I-1)+1 TO P(I)
```

In the expressions that change the subscript which identifies the country, in response to an arrow key press, the MOD function is used to wrap around from $P(I)$ to $P(I-1)+1$ and vice versa.

```
3760 J=((J-P(I-1)) MOD (P(I)-P(I-1))) + P(I-1)+1 : GOTO 3780
3770 J=((J-P(I-1)+P(I)-P(I-1)-2) MOD (P(I)-P(I-1))) + P(I-1)+1
```

Although each of these statements is longer than the combination of a simple assignment that increases or decreases J, followed by an IF statement that wraps it around if necessary, the expressions containing the MOD function are useful in programs that deal with circular queues (see Section 4.6).

```
1ØØ REM * MULDIREC uses a two-level directory *
1Ø1 :
11Ø  ON ERROR GOTO 8ØØØ                                        'set error trap
12Ø  KEY OFF: SCREEN 1                                       'set up screen
13Ø  COLOR 1,Ø: CLS
14Ø  DEF FN T$=CHR$(INT(256*RND))+CHR$(INT(256*RND))          'tiling function
15Ø  DIM U%(8ØØØ),V%(6ØØØ)                                 'arrays for bit maps
16Ø  XL=138                                           'x coord of country list
17Ø  XR=228                                    'x coord of schematic town list
18Ø  AU$=CHR$(24): AD$=CHR$(25}                        'up and down arrows
19Ø  AR$=CHR$(26): AL$=CHR$(27)                       'right and left arrows
299 :
3ØØ REM * transfer data to arrays *
31Ø  GOSUB 4ØØØ                                               'help file
32Ø  PRINT SPACE$(4Ø) "Please wait while data are loaded -";
33Ø   VC=CSRLIN                                          'cursor position
34Ø   HC=POS(Ø)                                           'of countdown
35Ø  I=Ø: J=Ø: K=Ø
36Ø  ON ERROR GOTO 41Ø                               'prepare to count items
37Ø  READ A$                                                  'read an item
38Ø   L$=LEFT$(A$,1)                          'increase the appropriate count
39Ø   IF L$="$" THEN I=I+1
              ELSE IF L$="#"
                 THEN J=J+1: LOCATE VC,HC: PRINT J ELSE K=K+1
4ØØ   GOTO 37Ø
41Ø  IF ERR<>4 OR ERL<>37Ø THEN 8ØØØ               'branch on an actual error
42Ø  RESUME 43Ø                                    'trap the end of data
43Ø  ON ERROR GOTO 72Ø                             'prepare to transfer
44Ø  IMAX=I: JMAX=J: KMAX=K
45Ø  DIM CONTINENT$(IMAX), COUNTRY$(JMAX), TOWN$(KMAX)        'allocate space
46Ø  DIM P(IMAX), Q(JMAX), YL(IMAX), YR1(JMAX), YR2(JMAX)       'for arrays
499 :
5ØØ REM * second pass
51Ø  RESTORE                                            'restart the data
52Ø  I=Ø: J=Ø: K=Ø
53Ø   READ A$
54Ø   L$=LEFT$(A$,1)                                    'extract left byte
55Ø   IF L$<>"$" THEN 6ØØ
56Ø    I=I+1                                                   'continent
57Ø    CONTINENT$(I)=MID$(A$,2)                       'store a continent
58Ø    P(I)=P(I-1)                           'initialize pointer to country list
59Ø    GOTO 53Ø
6ØØ   IF L$<>"#" THEN 68Ø
61Ø    J=J+1                                                  'country
62Ø    COUNTRY$(J)=MID$(A$,2)                           'store a country
63Ø    Q(J)=Q(J-1)                           'initialize pointer to town list
64Ø    P(I)=P(I)+1                           'increase pointer to country list
65Ø    LOCATE VC,HC
66Ø    PRINT JMAX-J                                    'display the countdown
67Ø    GOTO 53Ø
68Ø    K=K+1                                                   'town
69Ø    TOWN$(K)=A$                                      'store a town
7ØØ    Q(J)=Q(J)+1                           'increase pointer to town list
71Ø    GOTO 53Ø
72Ø  IF ERR<>4 OR ERL<>53Ø THEN 8ØØØ              'trap the end of data
73Ø  RESUME 74Ø
74Ø  ON ERROR GOTO 8ØØØ                              'reset the error trap
75Ø  PRINT SPACE$(4Ø) "Press the space bar to start."
76Ø  Z$=INPUT$(1)
799 :
```

```
800 REM * schematic of data base *
810  CLS
820   YL(0)=18                                      'top of schematic country list
830   YR1(0)=18                                      'top of its first section
840   YR2(0)=18                                      'top of schematic town list
850   FOR I=1 TO IMAX
860    YL(I)=18+172*P(I)/P(IMAX)          'top of section for Ith continent
870    FOR J=P(I-1)+1 TO P(I)
880     YR1(J)=YR1(J-1)+172/JMAX           'top of Jth section in country
890     YR2(J)=18+172*Q(J)/Q(JMAX)             'and in town schematics
900    NEXT J,I
910  LOCATE 1,1                                     'display the headings
920   PRINT " Continent      Country      Town"
930   PRINT " list           list         list"
940 GOSUB 6000
999 :
1000 REM * prompt *
1010  GOSUB 5200                                     'prompt for continent
1020   INPUT; "name of continent: ",B$
1030    IF B$<>SPACE$(LEN(B$)) THEN 1070            'test for default
1040     I=1                                           '1st continent
1050     PRINT CONTINENT$(I);
1060     GOTO 2000
1070   FOR I=1 TO IMAX                    'search for requested continent
1080    IF B$=CONTINENT$(I) THEN 2000
1090    IF B$<CONTINENT$(I) THEN 1110
1100   NEXT I
1110    Z$=B$: GOSUB 5000                                'name not found
1120   GOTO 1010
1999 :
2000 REM * a continent has been specified *
2010  GOSUB 5200
2020   PRINT CONTINENT$(I);                            'display the continent
2030  GOSUB 7000                            'display lower level directory
2040   GOSUB 5200
2050    IF INKEY$<>"" THEN 2050                     'clear keyboard buffer
2060    PRINT CONTINENT$(I) " - press arrow key (" AL$ AD$ AU$ AR$ ")";
2090    R$=INKEY$
2100     IF R$="" THEN 2090
2199 :
2200 REM * branch on arrow code *
2210  IF LEN(R$)<>2 THEN 2040                          'retry if needed
2220  ON INSTR("KPHM",RIGHT$(R$,1)) GOTO 2250,2270,2280,3000    'branch on arrow
2230   GOSUB 5000                                       'error branch
2240   GOTO 2040
2250  PUT (39,17),U%,PSET                        'restore primary display
2260   GOTO 1000
2270  I=(I MOD IMAX)+1: GOTO 2290               'select next continent
2280  I=((I+IMAX-2) MOD IMAX)+1                 'select prior continent
2290   PUT (39,17),U%,PSET                      'restore primary display
2300   GOTO 2000                               'display new country list
2999 :
3000 REM * prompt for a country *
3010  GOSUB 5200                                     'prompt for country
3020   INPUT; "name of country: ",C$
3030    IF C$<>SPACE$(LEN(C$)) THEN 3100            'test for default
3040     J=P(I-1)+1                           '1st country in continent
3050     GOSUB 5200
3060     PRINT COUNTRY$(J);
3070     GOTO 3210
```

```
3100    FOR J=P(I-1)+1 TO P(I)                              'search for country
3110     IF C$=COUNTRY$(J) THEN 3210                               'found
3120      IF C$<COUNTRY$(J) THEN 3140
3130    NEXT J
3140    Z$=C$: GOSUB 5000                                       'not found
3150    GOTO 2040                                               'try again
3199 :
3200 REM * country has been found *
3210    WH=Q(J)-Q(J-1)                                       'window depth
3220     IF WH>15 THEN WH=15                                 'limit to 15
3230    GET (157,61)-(319,65+8*WH),V%             'preserve the undercoat
3240    VIEW (158,62)-(318,64+8*WH),0,3             'create the window
3250    FOR KLO=Q(J-1)+1 TO Q(J) STEP 15          'display groups of names
3260      KHI=KLO+14
3270       IF KHI>Q(J) THEN KHI=Q(J)                      'final or only group
3280      LOCATE 9,21
3290      FOR K=KLO TO KHI                            'display single group
3300       PRINT USING "### \          \"; K,TOWN$(K)   'display single entry
3310       LOCATE CSRLIN,21                                  'reposition
3320      NEXT K
3330      IF KHI=Q(J) THEN 3390
3340      GOSUB 5200
3350      PRINT "press space bar for more";         'prompt for next group
3360        Z$=INPUT$(1)
3370        CLS                                      'clear the window
3380    NEXT KLO
3390    VIEW
3599 :
3600 REM * prompt for a country *
3610    GOSUB 5200
3620     IF INKEY$<>"" THEN 3620                    'clear keyboard buffer
3630      PRINT COUNTRY$(J) " - press arrow key (" AL$ AD$ AU$ ") or Q";  'prompt
3660    R$=INKEY$                                    'store key press code
3670     IF R$="" THEN 3660              •
3680     IF R$="Q" OR R$="q" THEN 10000                             'quit
3690     IF LEN(R$)<>2 THEN 3610                     'retry if not arrow
3700     ON INSTR("KPH",RIGHT$(R$,1)) GOTO 3720,3760,3770      'branch on arrow
3710      GOTO 3610                                  'retry if right arrow
3720      PUT (157,61),V%,PSET                    'revert to continent level
3730      GOTO 2040
3760      J=((J-P(I-1)) MOD (P(I)-P(I-1)))+P(I-1)+1:
                                                GOTO 3780      'next country
3770      J=((J-P(I-1)+P(I)-P(I-1)-2) MOD (P(I)-P(I-1)))
                                                +P(I-1)+1      'prior country
3780      PUT (157,61),V%,PSET                    'restore undercoat
3790      GOTO 3210
3999 :
4000 REM * explanation of prompts *
4010    MA$(1)=" to recall the continent list,"         'arrow roles at
4020    MA$(2)=" for the next continent,"                 'upper level
4030    MA$(3)=" for the previous continent,"
4040    MA$(4)=" to specify a country."
4050    MB$(1)=" to recall the country list,"           'arrow roles at
4060    MB$(2)=" for the next country,"                   'lower level
4070    MB$(3)=" for the previous country,"
4100    LOCATE 2,1                                  'explain the prompts
4110    PRINT "The program first prompts for the name"
4120    PRINT "of a continent. After you respond, it"
4130    PRINT "prompts for an arrow key (" AL$ AD$ AU$ AR$ ")."
4140    PRINT: PRINT "You press"      'explain arrow roles at upper level
```

```
4150    PRINT " " AL$ MA$(1)
4160    PRINT " " AD$ MA$(2)
4170    PRINT " " AU$ MA$(3)
4180    PRINT " " AR$ MA$(4)
4190   PRINT: PRINT "After you specify a country, the program";
4200   PRINT "prompts for an arrow key (" AL$ AD$ AU$ ") or Q."
4210   PRINT: PRINT "You press "              'explain arrow roles at lower level
4220    PRINT " "  AL$ MB$(1)
4230    PRINT " "  AD$ MB$(2)
4240    PRINT " "  AU$ MB$(3)
4250    PRINT " Q to quit."
4260   RETURN
4999 :
5000 REM * invalid key *
5010   GOSUB 5200
5020    PRINT Z$ " not found - press space bar";                      'diagnostic
5030    Z$=INPUT$(1)                                            'wait for key press
5040   RETURN
5200 REM * clear prompt area *
5210   LOCATE 25,1                                              'clear prompt area
5220    PRINT SPC(39)
5230   LOCATE 25,1
5240   RETURN
5399 :
6000 REM * upper level directory and associated schematics *
6010   VIEW(0,18)-(319,190)
6020    CLS                                              'clear the display area
6030    VIEW
6100    FOR I=1 TO IMAX                                  'loop through continents
6110     LOCATE 2*I+6,2
6120      PRINT CONTINENT$(I) TAB(13)                             'display continent
6130      PRINT USING "##"; P(I)                                  'and pointer
6140     LINE (8*POS(0)-6,8*CSRLIN-4)-(XL-2,YL(I))                    'tie line
6150     LINE (0,8*(2*I+5)-6)-STEP(112,16),3,B                    'bottom of box
6160     LINE (88,8*(2*I+5)-6)-STEP(0,16),3                      'and side box
6170     LINE (XL,YL(I))-(XL+30,YL(I-1)),3,B                    'box for schematic
6180      IF YL(I)-YL(I-1)>1
                THEN PAINT (XL+10,YL(I-1)+1),1+(I+1) MOD 3,3              'paint
6190     LINE (XL+32,YL(I))-(XL+35,YL(I-1)),3,B               'pointer depiction
6200      IF YL(I)-YL(I-1)>1
                THEN PAINT (XL+33,YL(I-1)+1),1+(I+1) MOD 3,3              'paint
6300      FOR J=P(I-1)+1 TO P(I)                                 'country loop
6310       IF J=P(I) THEN 6340                           'bypass in final cycle
6320       LINE (XL+1,YR1(J))-STEP(28,0),0          'bottom of box and side box
6330       LINE (XL+33,YR1(J))-STEP(2,0),0                'in country schematic
6340       LINE (XL+37,YR1(J))-(XR-2,YR2(J)),3                      'tie line
6350       LINE (XR,YR2(J))-(XR+35,YR2(J-1)),3,B        'box in town schematic
6360        IF YR2(J)-YR2(J-1)>1
                 THEN PAINT (XR+10,YR2(J-1)+1),FNT$,3                  'tile it
6370    NEXT J,I
6380    RETURN
6999 :
7000 REM * lower level directory for specified continent *
7010   GET (39,17)-(319,190),U%                             'preserve undercoat
7020    VIEW (XL+36,18)-(319,190)                         'clear tie line area
7030    CLS
7040    WH=P(I)-P(I-1)                                          'window height
7050     IF WH>13 THEN WH=13                                    'limit to 13
7060    VIEW (40,80)-(184,80+8*WH),0,3              'windows for country names
7070    VIEW (XR-3,30)-(XR+55,160),0,3           'and exploded town schematic
```

```
7Ø8Ø   VIEW
7Ø9Ø   PSW=Ø                                                 'switch for 1st group
71ØØ   YCP=3Ø                                      'top of 1st section in town schematic
72ØØ    FOR JLO=P(I-1)+1 TO P(I) STEP 13                  'loop through groups
721Ø     JHI=JLO+12
722Ø      IF JHI>P(I) THEN JHI=P(I)                          'final or only group
723Ø     LOCATE 11,7
73ØØ      FOR J=JLO TO JHI                                      'country loop
731Ø       PRINT USING "\          \###"; COUNTRY$(J),Q(J)   'name and pointer
732Ø       YCEXP=3Ø+13Ø*(Q(J)-Q(P(I-1)))/(Q(P(I))-Q(P(I-1)))  'bottom of section
733Ø       LINE (186,8*CSRLIN-12)-(XR-5,YCEXP),3                'tie line
734Ø        IF PSW OR YCEXP-YCP<4 OR J=JHI AND JHI<P(I)
                    THEN 738Ø                              'when appropriate
735Ø        LINE (XR-3,YCEXP)-STEP(58,Ø)                 'rule bottom of section
736Ø        PAINT (XR+1Ø,YCP+2),FNT$,3                           'and tile
737Ø        YCP=YCEXP
738Ø       LOCATE CSRLIN,7
739Ø      NEXT J
74ØØ     IF JHI=P(I) THEN 761Ø                     'jump if final or only group
741Ø     IF JHI>P(I-1)+13 THEN 749Ø             'jump unless 1st of several groups
742Ø      FOR J=JHI+1 TO P(I)
743Ø       YCEXP=3Ø+13Ø*(Q(J)-Q(P(I-1)))/(Q(P(I))-Q(P(I-1)))     'complete the
744Ø       LINE (XR-3,YCEXP)-STEP(58,Ø)                         'exploded town
745Ø       PAINT (XR+1Ø,YCP+2),FNT$,3                            'schematic
746Ø       YCP=YCEXP
747Ø      NEXT J
748Ø      PSW=1                            'switch shows 1st group was displayed
749Ø     GOSUB 52ØØ
75ØØ     IF INKEY$<>"" THEN 75ØØ                              'clear buffer
751Ø     PRINT CONTINENT$(I) " - press space bar for more";         'prompt
752Ø      R$=INKEY$
753Ø       IF R$="" THEN 752Ø                    'allow prompt for country
754Ø       IF LEN(R$)=2 THEN RETURN 222Ø         'if arrow key was pressed
755Ø      VIEW (4Ø,8Ø)-(184,8Ø+8*WH),Ø,3
756Ø      CLS                            'clear lower level directory area
757Ø      VIEW (186,18)-(XR-5,19Ø)
758Ø      CLS                                     'clear tie line area
759Ø      VIEW
76ØØ    NEXT JLO
761Ø    RETURN
7999  :
8ØØØ  REM * error trap *
8Ø1Ø   PRINT "Error" ERR "in line" ERL TAB(39)          'system error diagnostic
8Ø2Ø    RESUME 1ØØØØ
9999  :
1ØØØØ  REM * quit *
1ØØ1Ø   LOCATE 2Ø,1                                        'position Ok message
1ØØ2Ø   ON ERROR GOTO Ø                             'reset error trap address
1ØØ3Ø   END                                                       'quit
14999 :
15ØØØ  REM * lists of names *
15Ø1Ø  DATA $Asia
15Ø2Ø  DATA #Burma, Rangoon
15Ø3Ø  DATA #China, Canton, Changsha, Foochow, Nanking, Peking, Shanghai
15Ø4Ø  DATA #India, Bombay, Calcutta, Hyderabad, Madras, New Delhi
15Ø5Ø  DATA #Iran, Isfahan, Tabriz, Teheran
15Ø6Ø  DATA #Iraq, Baghdad, Basra
15Ø7Ø  DATA #Japan, Hiroshima, Kobe, Kyoto, Nagasaki, Osaka, Tokyo, Yamaguchi
15Ø8Ø  DATA #Pakistan, Amritsar, Islamabad, Karachi, Lahore
15Ø9Ø  DATA $Europe
```

```
15100 DATA #Albania, Tirana
15110 DATA #Austria, Vienna, Salzburg
15120 DATA #Belgium, Antwerp, Brussels, Liege, Ostend
15130 DATA #Bulgaria, Sofia
15140 DATA #Czechoslovakia, Bratislava, Prague
15150 DATA #Denmark, Copenhagen
15160 DATA #Finland, Helsinki
15170 DATA #France, Bordeaux, Lille, Lyons, Marseilles, Nice, Paris, Strasbourg
15180 DATA #Germany (DDR), Dresden, East Berlin
15190 DATA #Germany (FDR), Bonn, Frankfurt, Hamburg, Stuttgart, West Berlin
15200 DATA #Greece, Athens, Salonika
15210 DATA #Gt. Britain, Birmingham, Bristol, Cardiff, Coventry
15220 DATA Edinburgh, Glasgow, Liverpool, London, Manchester, Nottingham, York
15230 DATA #Hungary, Budapest, Szeged
15240 DATA #Ireland, Dublin
15250 DATA #Italy, Bologna, Florence, Genoa, Milan, Naples, Rome, Turin, Venice
15260 DATA #Netherlands, Amsterdam, Hague, Neijmagen, Rotterdam
15270 DATA #Norway, Bergen, Navarik, Trondheim, Oslo
15280 DATA #Poland, Gdansk, Lodz, Warsaw
15290 DATA #Portugal, Lisbon, Porto
15300 DATA #Rumania, Bucarest, Constanza
15310 DATA #Spain, Barcelona, Burgos, Madrid, Seville, Valencia
15320 DATA #Sweden, Gotheberg, Lund, Malmo, Stockholm, Uppsala
15330 DATA #U.S.S.R., Kiev, Leningrad, Moscow, Odessa
15340 DATA #Yugoslavia, Belgrade, Split
15350 DATA "$N. America"
15360 DATA #Canada, Montreal, Ottawa, Quebec, Toronto, Vancouver, Winnipeg
15370 DATA #Mexico, Chihuahua, Mexico City, Monterey, Tampico, Vera Cruz
15380 DATA #U.S.A., Atlanta, Baltimore, Birmingham, Boston, Charleston
15390 DATA Chicago, Cleveland, Dallas, Denver, Detroit, El Paso, Flint
15400 DATA Houston, Los Angeles, Miami, Milwaukee, Minneapolis, New Orleans
15410 DATA Phoenix, Portland, San Diego, San Francisco, Seattle, Washington
```

The towns in the Jth country are stored in slots $Q(J-1)+1$ through $Q(J)$ of the array TOWN\$. Thus, the number of text lines in the window that displays the towns in country J is set tentatively by:

```
3210 WH=Q(J)-Q(J-1)
```

The window height is reset to 15 if the tentative value is greater. Then the names of these towns are displayed in groups of 14 by the loop:

```
3250 FOR KLO=Q(J-1)+1 TO Q(J) STEP 15
...
3380 NEXT KLO
```

Most of the program deals with the visual effects that reinforce the basic idea of using a multilevel directory. They include:

1. Drawing the schematic lower-level directory (country list) and the schematic town list and drawing the tie lines

2. Displaying the portion of the lower-level directory for a particular continent and exploding the corresponding portion of the schematic town list

3. Preserving and restoring the portion of the display that (2) alters

4. Displaying the towns in a particular country

5. Preserving and restoring the portions of the display that (4) alters

Suggested Activities

1. Adapt the MULDIREC program to read data from a file on disk. Use the same method to identify data of different levels, and use a two-pass process to set up and fill the arrays. Detect the end of the file with the EOF function.

2. Adapt the MULDIREC program as follows. Omit the pointers, the schematics, and the tie lines from the display. Allow several further levels of directory. Use the right and left arrow keys to increase and decrease the level under attention. Display the data at successive levels of detail in windows that begin at intervals across the screen. Use the down and up arrow keys to pan within the most detailed level of data currently displayed. Use the PgDn and PgUp keys to move from group to group of entries within this body of data and the Home and End keys to move to its beginning and end, respectively.

3. Extend the adaptations in Activity 2 (*a*) to read data from a file on disk and (*b*) to look for items in each level of directory by a binary search when the length warrants it.

2.3   Merging a Pair of Lists

The primary purpose of the MERGER program is to merge, in alphabetic order, two files of data that have been saved on disk. As a preliminary, MERGER can be used to sort a list of words into alphabetic order and save the alphabetized list on disk. You type the words in response to a simple prompt, and you may use up to eight letters in the name you give to the file for future reference. After sorting and saving one list, you can sort and save another. When you have created two files of sorted data on disk in this way, you can use the program to merge the two files into a single file that is written on disk in alphabetic order, too. Then you can go on to merge files that have been formed by merging. The program can sort up to 100 words, but there is no limit to the size of the files that it will merge.

Merging thus complements sorting as an activity of immense importance in the maintenance of ordered lists and files. Some methods of sorting, in fact, depend on putting small batches of data into order and merging them into successively larger batches that also are in the required sequence.

The MERGER program begins by prompting you to type S to sort new input data, M to merge two sorted files, or Q to quit. If you just press the S key, the program prompts you to type the items and press the Enter key twice after the last item to start sorting. You could respond, for example, by typing the words fly, beetle, ant, and butterfly. When it is sorting, the program displays the word "Sorting" with a countdown from the number of items to 0 alongside. Then the program prompts for the name of the output file in which the sorted data will be saved. You respond by typing, for example, BUGS. The file is written, and the initial prompt returns. Now you can create a sorted file, called BIRDS, that consists of the names of several birds in a completely similar fashion.

To merge two files, you type M when the menu is on the screen. The program prompts for the names of the two input files and the output file. Suppose that you respond by typing BUGS, BIRDS, and BIRDSNBS, respectively. Suppose that the BUGS file consists of the items ant, beetle, butterfly, fly. Suppose, too, that the BIRDS file consists of the items auk, emu, jay, robin.

The action continues as follows. The program displays

```
Next item
in file 1
↓
ant─┐
    ├─>
auk─┘
↑
Next item
in file 2

Press space bar to compare
```

When you press the space bar, the word "ant" turns red to show that it has alphabetic precedence over auk. This is shown here by boldface type.

```
ant─┐
    ├─>
auk─┘
```

The prompt changes to tell you to move ant into the output file. Then the center of the display changes to

```
ant┐
───┤─> ant
auk┘
```

The word "ant" at the left of the screen has turned light green to show it has been used. This is shown here by underscoring. Then the center part of the display changes to

```
beetle┐
      ├─> ant
auk───┘
```

   The program continues to prompt you to compare items, to move the appropriate item into the output stream, and to input the next item from file 1 or file 2 in turn. After successive key presses, the word "auk" turns red to show that it has alphabetic precedence over beetle. Then auk is moved to the output file and the next item (emu) is brought from file 2 for comparison.

```
beetle┐              beetle┐               beetle┐
      ├─> ant              ├─> auk  ant           ├─> auk  ant
auk───┘              auk───┘               emu───┘
```

The word "beetle" turns red because it has precedence over emu, and beetle is moved into the output.

```
beetle┐                    beetle┐
      ├─> auk  ant               ├─> beetle  auk  ant
emu───┘                    emu───┘
```

When you press the space bar in response to the next few prompts, the center part of the display changes in succession to

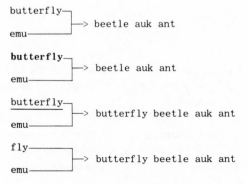

```
butterfly┐
         ├─> beetle  auk  ant
emu──────┘

butterfly┐
         ├─> beetle  auk  ant
emu──────┘

butterfly┐
         ├─> butterfly  beetle  auk  ant
emu──────┘

fly┐
   ├─> butterfly  beetle  auk  ant
emu┘
```

```
fly
        ─┐
         ├─> butterfly beetle auk ant
emu     ─┘

fly     ─┐
         ├─> emu butterfly beetle auk ant
emu     ─┘

fly     ─┐
         ├─> emu butterfly beetle auk ant
jay     ─┘

fly     ─┐
         ├─> emu butterfly beetle auk ant
jay     ─┘

fly     ─┐
         ├─> fly emu butterfly beetle auk ant
jay     ─┘
```

Each group of three displays shows that (1) a new item has been brought in for comparison, (2) the item that has alphabetic precedence has been emphasized, and (3) the item with precedence has been moved into the output stream and deemphasized at the left of the screen as a preliminary to being replaced by the next item in the file that it came from. When you press the space bar again, the display changes to:

```
        ┌─> fly emu butterfly beetle auk ant
jay    ─┘
```

The prompt tells you that all the items in file 1 have been used and to run out file 2. This happens because fly is the final item in the BUGS file. By coincidence, the word "ant" just reaches the right edge of the screen at this time. A few more key presses complete the merge.

. . .

```
        ┌─> jay fly emu butterfly beetle auk
robin  ─┘

        ┌─> robin jay fly emu butterfly beetl
robin  ─┘
```

The word "robin" at the left of the screen has turned light green to show that it has been used, and the word "beetle" is truncated at the right of the screen. Pressing the space bar brings back the main menu.

MERGER uses the following very simple algorithm. We use the word "comparand" for any item that is compared with another to determine which comes first alphabetically or numerically or in accordance with whatever other criterion was used to sort the lists being merged.

1. Make the first item in file 1 the first comparand.

2. Make the first item in file 2 the second comparand.

3. If (*a*) the first comparand comes before the second comparand in alphabetic order or (*b*) the comparands are the same, then (*c*) copy the first comparand into the output file and (*d*) fetch the next item from file 1 for use as the first comparand. Otherwise, (*e*) copy the second comparand into the output file, and (*f*) fetch the next item from file 2 for use as the second comparand.

4. Repeat step 3 unless the comparand which was copied into the output file ended the file that it came from. In that case perform step 5.

5. Copy the other comparand into the output followed by all the remaining entries in the file that it came from.

This statement tacitly assumes that neither input file is null, that is, empty. The algorithm can be restated to cover the cases that one or both files are empty, as follows:

1. Fetch the first item from file 1 for use as the first comparand and go on to step 2 unless file 1 is empty. In the latter case make the output file a copy of file 2 and stop.

2. Fetch the first item from file 2 for use as the second comparand and go on to step 3 unless file 2 is empty. In the latter case make the output file a copy of file 1 and stop.

Steps 3 to 5 are the same as steps 3 to 5 of the preceding statement of the algorithm that assumes neither file is null.

You can see immediately that this second algorithm works when file 1 is null but file 2 is not null and in the opposite case also. When both files are null, the fact that file 1 is null leads step 1 to make the output file a copy of file 2. This makes the output file null too, because file 2 also is null.

The algorithm allows repeated items; that is, the same item is repeated in the output file when it occurs in both input files. Depending on circumstances, you may want to allow repetitions or you may want to alter the program to throw them away. In the second of these circumstances, you may also want to discard redundant items brought together in the sorting step and keep a record of when redundancies occur and when they are discarded.

Although the algorithm is simple, care must be taken to anticipate different conditions that can arise in its execution and to design test data that gives rise to these conditions. Thus, the following cases should be included:

1. Both files are null.

2. One file is null, and the other file is not.

3. All the entries in one file come before the first entry in the other file, alphabetically.

4. A portion of the merged file consists of elements taken in alternation from the two input files.

5. The contiguity of two or more elements of one input file is preserved in the output file.

In each of the cases 2, 3, and 5, the two files have different roles. Each case should be tested with input file 1 serving one role and input file 2 serving the other; for example, file 1 is null and file 2 is not. Each case should then be tested with the roles of the two files reversed. This can be done in separate tests that use the same pair of files, in one test using them as files 1 and 2 and in the other test using them as files 2 and 1, respectively.

Much of the program relates to the graphical display of the ongoing process. The horizontal tie lines extend initially from the two comparands to a position two places to the right of the longer. The tie lines are not changed when the first item is moved into the output line. When the first comparand that has been put into the output is replenished, the tie lines are extended if the replenishment is longer than both (1) its predecessor and (2) the residual comparand, that is, the comparand retained from the preceding comparison. This occurred earlier when beetle succeeded ant. In that circumstance, the tie lines were extended two places to the right of the replenishment and the output line was moved to the right correspondingly.

When the replenishment comparand is shorter than both the comparands that were on the screen previously (for example, when fly succeeded butterfly), the tie lines are not shortened. In that circumstance, the rightmost ends of the tie lines are kept unchanged, because shortening them would move the output line backwards on the screen, which we think is undesirable visually. In the next cycle that moves a comparand into the output stream, the tie lines are shortened to keep the display compact. This shortening can make the words already in the output stream move to the left, which happened, in fact, when fly was moved into the output. This is not so undesirable visually.

Program Variables

ILIM	The limit to the number of items that can be sorted
A$(i)	The ith item in the list that is to be sorted or that has been sorted
P(j)	The jth element in the list of pointers to the array A$

R$	The one-character response to certain prompts
I	The index of items in the loop that lets you type a list of data
J,K	The indexes of the outer and inner loops of the sorting process; this applies the insertion sort algorithm to a pointer list
PC	The pointer to the item for which the correct position among its predecessors is sought in the current cycle of the sorting process
AC$	The actual item
NF$	The name of the output file used to save data that has been sorted
INF1$	The name of the first input file in a merge operation
INF2$	The name of the second file
OUTF$	The name of the merged file
FFN	The flag that shows the file or files from which data should be read (0 initially to read an item from each file, 1 to read the next item from the first file, 2 to read the next item from the second file)
M1	The flag that is 1 initially and 0 when the end of file 1 has been reached
M2	The corresponding flag for file 2
S$	The portion of the output file currently displayed on the screen
HC	The horizontal cursor position of the right end of the tie lines drawn from the comparands currently on the screen
HCOLD	The value of HC in the preceding cycle that compared items from the two files
HC1	The cursor position of the start of the tie line from the first comparand
HC2	The corresponding value for the second comparand
HCNEW	The value used for HC immediately after the transfer of an input item to the output
B$	The current comparand from the first file
C$	The current comparand from the second file
D$	The comparand selected for output

Program Notes

The bulk of the merging process is performed by the WHILE . . . WEND loop with the overall control structure:

```
3010 WHILE M1 AND M2
3020   IF FFN=2 THEN 3070
3030     IF EOF(1) THEN M1=0: GOTO 3070
3040     INPUT #1,B$
. . .
```

```
3070   IF FFN=1 THEN 3120
3080    IF EOF(2) THEN M2=0;  GOTO 3120
3090     INPUT #2,C$
3120   IF M1=0 OR M2=0 THEN 3730
. . .
3350   IF B$<=C$ THEN D$=B$ ELSE D$=C$
3360    FFN=2+(B$<=C$)
. . .
3730 WEND
```

The flags M1 and M2 are 1 initially, and FFN is 0 initially. FFN is changed to 1 and 2 within the loop when the item from the first or second file, respectively, is transferred to the output. The statements in the WHILE . . . WEND loop thus:

1. Input items from both files at the start of the merge
2. Replenish the appropriate comparand in later cycles
3. Deal with the cases in which file 1 ends before file 2 or file 2 ends before file 1 and neither file is null
4. Deal correctly with the cases in which (a) only file 1 is null, (b) only file 2 is null, and (c) both files are null

Two more WHILE . . . WEND loops handle the run-out actions.

Suggested Activities

1. Alter the program to suppress repeated occurrences of items in files produced (a) by sorting and (b) by merging. [*Hint:* To take care of (b), expand line 3360 to set FFN to 0 when the comparands are equal.]

2. Alter the program to include repeated items just once in files produced by sorting and merging but include the cumulative numbers of occurrences in parentheses. Thus, if "fly" occurs twice in the sorted data, the program will include "fly (2)" in the file that it writes to disk. If "fly (2)" and "fly (3)" occur in two files that are merged, the program writes "fly (5)" in the output file. This kind of action is used, for example, to find the frequency with which words occur in a body of text.

3. Alter the program to use the medium-resolution graphics mode. Draw the tie lines with LINE statements.

4. Enumerate different situations that can occur in the execution of the program and construct data to test them.

```
100 REM * MERGER merges two files *
101 :
110   ON ERROR GOTO 7000                                    'set error trap
120     KEY OFF: SCREEN 0: WIDTH 40              'set up the screen
130     COLOR 7,1,1: CLS
140     ILIM=100                                            'sorting limit
150     DIM A$(ILIM),P(ILIM)                          'arrays for sorting
199 :
200 REM * initial prompt *
210   CLS
220   LOCATE 6,5                                                'main menu
230     PRINT "Please type:"
240     PRINT SPACE$(47) "S to sort new input data"
250     PRINT SPACE$(47) "M to merge two sorted files"
260     PRINT SPACE$(47) "Q to quit: ";
270   R$=INPUT$(1)
280   ON INT((1+INSTR("sSmMqQ",R$))/2) GOTO 1000,2010,8000   'store the response
290     PRINT SPACE$(87) "Invalid response"                         'branch on it
300     PRINT SPACE$(47) "press space bar and try again"
310     Z$=INPUT$(1)
320   GOTO 210                                  'retry after invalid response
999 :
1000 REM * type unsorted data, sort and save it *
1010  LOCATE 14,8                                            'instructions to
1020    PRINT "Please press the Enter key"                      'type data
1030    PRINT SPACE$(47) "after each item and twice"          'for sorting
1040    PRINT SPACE$(47) "after the last item to"
1050    PRINT SPACE$(47) "start sorting" SPACE$(40)
1060    I=1                                       'initialize the item count
1070    LOCATE CSRLIN,14                            'position the cursor
1080    LINE INPUT A$(I)                                'store an item
1090    WHILE LEFT$(A$(I),1)=" "                   'remove leading spaces
1100      A$(I)=MID$(A$(I),2)
1110    WEND
1120    IF A$(I)="" THEN I=I-1: GOTO 1200                 'start the sort
1130    IF I<ILIM THEN I=I+1: GOTO 1070          'loop back for another item
1140    PRINT SPC(40) "This list will be sorted"
1150    PRINT "when you press the space bar"           'limit has been reached
1160    Z$=INPUT$(1)
1199 :
1200 REM * sort *
1210  CLS
1220  LOCATE 10,5                                         'display message
1230    PRINT "Sorting" I
1240  FOR J=1 TO I                              'set up the pointer list
1250    P(J)=J
1260  NEXT J
1270  FOR J=2 TO I                              'outer cycle of insertion sort
1280    LOCATE 10,13
1290    PRINT I-J SPC(2)                                    'countdown
1300    PC=P(J)                              'pointer to item being placed
1310    AC$=A$(PC)                             'actual item being placed
1320    K=J                                    'position in pointer list
1330    WHILE AC$<A$(P(K-1)) AND K>1                       'inner cycle
1340      P(K)=P(K-1)                              'move pointer down
1350      K=K-1                             'decrease the indirect pointer
1360    WEND
1370    IF K<J THEN P(K)=PC          'move the original pointer if necessary
1380  NEXT J
1390  LOCATE 10,13
```

```
1400    PRINT "is complete"
1499 :
1500 REM * save sorted file *
1510    PRINT                                          'prompt for file name
1520    INPUT "    Name of output file: ",NF$
1530    OPEN NF$ FOR OUTPUT AS #1                          'open the file
1540    FOR J=1 TO I
1550      PRINT TAB(14) A$(P(J))
1560      PRINT #1,A$(P(J))                               'write it to disk
1570    NEXT J
1580    CLOSE #1                                          'close the file
1590    GOTO 200
1999 :
2000 REM * merge two files *
2010    CLS
2020    LOCATE 10,1                                    'prompt for file 1
2030    INPUT "Name of 1st input file: ",INF1$
2040    PRINT                                          'prompt for file 2
2050    INPUT "Name of 2nd input file: ",INF2$
2060    PRINT                                          'prompt for output file
2070    INPUT "Name of output file:    ",OUTF$
2080    OPEN INF1$ FOR INPUT AS #1                        'open the files
2090    OPEN INF2$ FOR INPUT AS #2
2100    OPEN OUTF$ FOR OUTPUT AS #3
2110    CLS                                            'clear the screen
2120    LOCATE 7,1                                    'label item from file 1
2130      PRINT "Next item"
2140      PRINT "in file 1"
2150      PRINT " " CHR$(25)
2160    LOCATE 13,1                                   'label item from file 2
2170      PRINT " " CHR$(24)
2180      PRINT "Next item"
2190      PRINT "in file 2"
2200    M1=1                                          'set file 1 flag
2210    M2=1                                          'set file 2 flag
2220    FFN=0                          'set switch to fetch from both files
2230    S$=""                        'initialize the output display string
2240    HC=0                          'the join position on the tie lines
2250    HCOLD=0                                    'and its heldover value
2999 :
3000 REM * main merging cycle *
3010    WHILE M1 AND M2                        'cycle while both files are active
3020      IF FFN=2 THEN 3070            'jump unless item needed from file 1
3030        IF EOF(1) THEN M1=0: GOTO 3070           'jump if file 1 ended
3040        INPUT #1,B$                      'fetch next item from file 1
3050          LOCATE 10,1                                    'display it
3060          PRINT B$
3070      IF FFN=1 THEN 3120            'jump unless item needed from file 2
3080        IF EOF(2) THEN M2=0: GOTO 3120           'jump if file 1 ended
3090        INPUT #2,C$                      'fetch next item from file 2
3100          LOCATE 12,1                                    'display it
3110          PRINT C$
3120      IF M1=0 OR M2=0 THEN 3730      'jump if either file has ended
3130      HC1=LEN(B$)+1                          'set lengths of comparands
3140      HC2=LEN(C$)+1
3150      IF HC1<HC2 THEN HCNEW=HC2 ELSE HCNEW=HC1           'set join position
3160      IF HCNEW>HC THEN HC=HCNEW
3170      GOSUB 6000                                    'draw the tie line
3180      HCOLD=HC                                    'preserve join position
3190      HC=HCNEW
```

```
3299 :
3300 REM * compare *
3310   LOCATE 17,1                                                      'prompt
3320   PRINT "Press space bar to compare"
3330   PRINT SPACE$(80)                              'clear rest of message area
3340   Z$=INPUT$(1)
3350   IF B$<=C$ THEN D$=B$ ELSE D$=C$                  'select item for output
3360     FFN=2+(B$<=C$)                                'set replenishment switch
3370     LOCATE 8+2*FFN,1                    'vertical text position of comparand
3380     COLOR 4
3390     PRINT D$                                        'redisplay item in red
3499 :
3500 REM * move selected item to output file *
3510   COLOR 7                                          'switch back to white
3520   LOCATE 17,1
3530   PRINT "Press space bar to" TAB(40)                               'prompt
3540   PRINT "move " D$ TAB(40)
3550   PRINT "into output file" TAB(40)
3560   Z$=INPUT$(1)
3570   COLOR 3                                        'redisplay item in green
3580   LOCATE 8+2*FFN,1
3590   PRINT D$
3600   COLOR 7                                          'switch back to white
3610   GOSUB 6000                                        'draw the tie line
3620   S$=LEFT$(D$+" "+S$,40)               'update the output display string
3630   LOCATE 11,HC+5
3640     PRINT LEFT$(S$,35-HC) TAB(40)                          'redisplay it
3650   HCOLD=HC                                    'update the join position
3660   HC=HCNEW
3670   PRINT #3, D$                          'write the selected item to disk
3680   LOCATE 17,1
3690     PRINT "Press space bar to" TAB(40)                             'prompt
3700     PRINT "input next item from" TAB(40)
3710     PRINT "file" FFN TAB(40)
3720   Z$=INPUT$(1)                                       'wait for key press
3730 WEND
3999 :
4000 REM * run out condition *
4010   IF M2 THEN 4020 ELSE 4500               'branch if file 1 still active
4020   LOCATE 17,1
4030     PRINT "All the items in file 1" TAB(40)      'prompt to run out file 2
4040     PRINT "have been used. Press" TAB(40)
4050     PRINT "the space bar to run" TAB(40)
4060     PRINT "out file 2." TAB(40)
4070   LOCATE 10,1                              'clear final item of file 1
4080     PRINT TAB(40)
4090   WHILE M2                              'cycle while file 2 is active
4100   HC2=LEN(C$)+1
4110     IF HC<HC2 THEN HC=HC2                          'set join position
4120   LOCATE 12,HC2
4130     GOSUB 4400
4140     IF HC>HCOLD THEN PRINT LEFT$(S$,41-POS(0))     'redisplay output line
4150   HC=HC2
4160   Z$=INPUT$(1)
4170   COLOR 4                                      'redisplay item in red
4180   LOCATE 12,1
4190     PRINT C$;
4200   Z$=INPUT$(1)
4210   COLOR 3                                      'redisplay item in green
4220   LOCATE 12,1
```

```
423Ø     PRINT C$;
424Ø     GOSUB 44ØØ                                      'redraw the tie line
425Ø      S$=C$+" "+S$                               'update the output string
426Ø     LOCATE 11,HC+5
427Ø      PRINT LEFT$(S$,41-POS(Ø))                             'redisplay it
428Ø     PRINT #3,C$                                         'write to disk
429Ø     HCOLD=HC                                   'preserve tie line length
43ØØ     Z$=INPUT$(1)
431Ø     IF EOF(2) THEN M2=Ø: GOTO 435Ø        'branch if file 2 is exhausted
432Ø     INPUT #2,C$                                    'fetch the next item
433Ø     LOCATE 12,1
434Ø      PRINT C$;                                             'redisplay it
435Ø     WEND
436Ø     GOTO 5ØØØ
44ØØ REM * redraw the tie line *
441Ø     COLOR 7                                    'redraw start of tie line
442Ø      PRINT STRING$(HC-HC2+1,196) CHR$(217) TAB(4Ø)
443Ø     LOCATE 11,1
444Ø      PRINT SPC(HC) CHR$(218) STRING$(2,196) ">";      'redraw rest of it
445Ø     RETURN
4499 :
45ØØ REM * run out file 1 *
451Ø     IF M1 THEN 452Ø ELSE 5ØØØ               'branch if file 2 still active
452Ø     LOCATE 17,1
453Ø      PRINT "All the items in file 2" TAB(4Ø)                     'prompt
454Ø     PRINT "have been used. Press" TAB(4Ø)
455Ø     PRINT "the space bar to run" TAB(4Ø)
456Ø     PRINT "out file 1." TAB(4Ø)
457Ø     LOCATE 12,1                              'clear final item of file 2
458Ø      PRINT TAB(4Ø)
459Ø     WHILE M1                                 'cycle while file 1 is active
46ØØ      HC1=LEN(B$)+1                            'length of current item
461Ø     IF HC<HC1 THEN HC=HC1                         'set join position
462Ø     LOCATE 1Ø,HC1
463Ø      GOSUB 49ØØ                                   'redraw the tie line
464Ø      IF HC>HCOLD THEN PRINT LEFT$(S$,41-POS(Ø));   'redisplay output line
465Ø     HC=HC1
466Ø      Z$=INPUT$(1)
467Ø     COLOR 4                                  'redisplay the item in red
468Ø     LOCATE 1Ø,1
469Ø      PRINT B$;
47ØØ      Z$=INPUT$(1)
471Ø     COLOR 3                                 'redisplay the item in green
472Ø     LOCATE 1Ø,1
473Ø      PRINT B$;
474Ø     GOSUB 49ØØ                                    'redraw the tie line
475Ø      S$=B$+" "+S$                             'update the output string
476Ø     LOCATE 11,HC+5
477Ø      PRINT LEFT$(S$,41-POS(Ø));                   'redisplay output line
478Ø     PRINT #3,B$                                   'write the item to disk
479Ø     HCOLD=HC                                  'preserve tie line length
48ØØ      Z$=INPUT$(1)
48Ø5     IF EOF(1) THEN M1=Ø: GOTO 484Ø         'branch if file 1 is exhausted
481Ø     INPUT #1,B$                                      'fetch next item
482Ø     LOCATE 1Ø,1
483Ø      PRINT B$                                             'display it
484Ø     WEND
485Ø     GOTO 5ØØØ
49ØØ REM * redraw tie line *
491Ø     COLOR 7                                    'redraw start of tie line
```

```
492Ø    PRINT STRING$(HC-HC1+1,196) CHR$(191) TAB(4Ø)
493Ø    LOCATE 11,1
494Ø    PRINT SPC(HC) CHR$(192) STRING$(2,196) ">";          'redraw rest of it
495Ø    RETURN
4999 :
5ØØØ REM * merge is complete *
5Ø1Ø    COLOR 7
5Ø2Ø    LOCATE 17,1                                          'display message
5Ø3Ø    PRINT "Merge is complete -" TAB(4Ø)
5Ø4Ø    PRINT SPACE$(8Ø) "press space bar to continue" TAB(4Ø)
5Ø5Ø    PRINT: PRINT SPACE$(12Ø);
5Ø6Ø    CLOSE #1,2,3                                          'close all files
5Ø7Ø    Z$=INPUT$(1)                                          'wait for key press
5Ø8Ø    GOTO 2ØØ                                              'go back to main menu
5999 :
6ØØØ REM * draw the tie lines *
6Ø1Ø    LOCATE 1Ø,HC1                                         'draw upper part
6Ø2Ø    PRINT STRING$(HC-HC1+1,196) CHR$(191) TAB(4Ø)
6Ø3Ø    LOCATE 11,1                                           'draw center part
6Ø4Ø    PRINT SPC(HC) CHR$(195) STRING$(2,196) ">";
6Ø5Ø    IF HC>HCOLD THEN PRINT LEFT$(S$,4Ø-POS(Ø))            'redisplay output string
6Ø6Ø    LOCATE 12,HC2
6Ø7Ø    PRINT STRING$(HC-HC2+1,196) CHR$(217) TAB(4Ø)         'draw lower part
6Ø8Ø    RETURN
6999 :
7ØØØ REM * error trap *
7Ø1Ø    LOCATE 2Ø,1
7Ø2Ø    PRINT "Error" ERR "in line" ERL TAB(4Ø)              'diagnostic message
7Ø3Ø    PRINT SPACE$(8Ø) "Press space bar for main menu";
7Ø4Ø    Z$=INPUT$(1)
7Ø5Ø    RESUME 2ØØ                                            'go back to main menu
7999 :
8ØØØ REM * quit *
8Ø1Ø    CLOSE #1,2,3                                          'close all files
8Ø2Ø    COLOR 7,Ø
8Ø3Ø    WIDTH 8Ø
8Ø4Ø    LOCATE 1,1                                            'position the Ok message
8Ø5Ø    ON ERROR GOTO Ø                                       'reset error trap address
8Ø6Ø    END                                                  'quit
```

5. Expand the program to merge four files concurrently. (*Hint:* Save the command FILES = 10 in a file called CONFIG.SYS on the disk that you use to boot the PC. Then invoke BASICA by the DOS level statement BASICA/F:5 to allow five files to be open at the same time when you run the program.)

6. Write a program to merge a small file A into a large file B as follows. The program (*a*) reads, say, the first 1024 elements of file B into an array, (*b*) uses a binary search to find where the first element of A should be inserted, (*c*) outputs the elements of B that precede that point, (*d*) outputs the first element of A, (*e*) gets the next element of A, repeats the overall process, and replenishes the array with the next block of elements of B whenever appropriate. This technique reduces the number of comparisons considerably.

2.4 Some Further Directions

The principles that have been explained in Sections 2.1 to 2.3 can be extended to meet many practical needs that arise when files are searched and updated. This section discusses a few of those methods and shows how some of the ideas in the earlier sections lead to more general topics and relate to techniques that are described in other chapters.

The simplest extension of the searching process of Section 2.1 is illustrated by a file in which each record consists of the social security number and the date of birth of an employee of a company. Suppose that the file has been sorted on the social security number and that the data from the ith record is stored as the elements S(i) and D(i) of two arrays that are large enough to hold the entire file. Thus S(i) is the social security number of the ith employee, in the order of increasing social security number, and D(i) is the employee's birth date. If SK is the social security number of the employee whose birth date is needed, the methods of Section 2.1 can be used to find the index M of the item S(M) that is equal to SK. The required birth date is then D(M).

In the happy circumstance that all the data in the file can be stored in high-speed memory at the same time, the search is said to be internal. Most often, however, the number and size of the records and the limitations of high-speed memory make internal search impossible. Then the search must refer to data on disks or tapes or other auxiliary storage media, and it is said to be external.

If the file is stored on magnetic tape, the records are in a physical order that the user knows. The most primitive form of external search works through the records in this order serially. Each record is read in turn. The datum in the appropriate field is compared with the value that must be matched, and the search ends successfully when a match occurs or unsuccessfully when the end of the file is reached. If the file was presorted on the field that is used for searching and the record that is being sought is absent, then the search can be stopped as soon as the required value of the search key has been passed. Even then, however, a serial search that ties up the entire computer can be disastrously slow, particularly if the file is very long. Special attachments that transmit data at considerable speed through matching devices without delaying the other actions of the computer can make serial searching very effective. For the present, however, they are not standard features of everyday computing.

Quite often it is possible to store the values of the keys of all the records of a file in high-speed memory even though the complete records are too voluminous to store that way. A program then can perform a binary search on the list of keys stored in an internal array. The subscript of the item that matches the requisite value of the key is the serial number of the record to be retrieved from the file. It can be used in several ways. For example, if the system includes a command to skip rapidly over N records of an external file, then the retrieval program sets N to 1 less than the serial number of the required record, the skip command is executed, and the next record in the file is read. This kind of skipping is a standard tape operation on many computers. A similar physical process is performed in computer-controlled microfilm readers. Some systems also have a command to backspace N records. This expedites a succession of search requests that are not in serial order of the target records.

Most people who write applications programs that use magnetic disks do not try to take direct account of the specific disk addresses at which records are written and read. The information is used, however, by several programming resources for high-level languages that facilitate external searching. One common technique uses *indexed sequential files*. In a file of that kind, the records are sorted on a field that puts them in the order that is most convenient for serial processing. For example, a personnel file might be presorted on the employee social security number. The records and some supplementary information that constitute one or more indexes are stored on disk with two objectives in view.

The first objective is to allow serial processing when this is needed, for example, to issue paychecks to the entire staff. The second objective is to process individual records without working serially through the entire file, for example, to issue revised paychecks to just one or two employees. The file is treated as a series of pieces containing several records each. The ith entry in the index consists of (1) the value of the key in the first record of the ith piece and (2) the information that the system needs to find that piece of the file on the disk.

The index is searched sequentially if it is short, or by the methods of Section 2.1 if it is long, to find the record that corresponds to a particular value of the key. If the search key is less than the first entry in the index, the file does not contain a corresponding record. If the key matches an entry in the index, the required record is the first in the corresponding piece of the file. In general, if the search key falls between the ith entry in the index and the $(i+1)$th entry, the required record is in the ith piece if it is present at all.

Different systems then proceed in different ways. One tactic is to simply search the relevant piece of the file serially. Another tactic is to read a detailed index of the relevant piece of the file into high-speed memory and search it to find the precise disk address of the required record. Yet another tactic is to bring the entire piece of the file into high-speed memory for searching and possible alteration. When the file is very long, the index to successive pieces of a convenient size may be too large to fit into high-speed memory. Then the index itself can be treated as a succession of pieces and accessed via a second-level index. The indexes in this approach to searching are analogous to the directories in Section 2.2.

The indexed sequential approach helps when the datum that is used to sort the file is also used as the search key. Given the capability to find the record with a particular value of one key, however, a lot more flexibility is possible. For example, take a personnel file that has been sorted on social security numbers and form from it another file in which each record consists of (1) the social security number of the corresponding record of the first file and (2) the name of the person described by the record. Sort this second file to put the names in alphabetic order. The result is a file of names that can be searched efficiently to find the social security number of a person using his or her name as the key. Then the file can be used to retrieve the appropriate record by using the methods already described. The file that leads from the names to the social security numbers is also called an index. Similar methods are an important element of large-scale centralized information retrieval services that can be accessed from afar via networking and are used extensively by reference librarians.

We have dealt so far with processes that just retrieve information from a file during periods in which it stays unchanged. There are massive applications of this kind, such as information retrieval systems based on data updated only once a week or at even longer intervals. In this situation, new data can be merged into the files and the indexes reconstructed, perhaps at a time when the system can be made inaccessible to users without hardship. There are other applications, however, such as transaction processing, that require the instantaneous revision of selected records. Matters are simplified when the length of the record does not change and new data can be written over the old data in the record. The deletion of a record can then be handled in a simple way, too, just by changing the datum in a one-byte field kept as an indicator of whether the record is to be considered active or dead. Also, the corresponding entry in a detailed index can be flagged.

The situation is more difficult when individual records cannot be overwritten, when revision may increase the length of a record, and when new records may be added. A variety of techniques are used to cope with indexed sequential files and other kinds of file organization. One tactic puts new items into a supplementary file that is merged with the main file periodically. When the system searches for a record, it first checks the main file by using an efficient method. If the search is unsuccessful, the supplementary file is searched, in a cruder way if necessary, in case the request refers to a record that was added since the most recent merge.

A widely used technique is to allow extra space in each of the blocks into which the file is divided. A new record is put into the extra space in the appropriate block. Further measures (1) allow separate consideration of the additional records in later searches or (2) integrate each record when it is added to the block in a way that shows its logical position relative to the other records. The linked list techniques of Chapter 3 are of paramount importance in this approach. Periodically, each block is tidied by a program that makes the physical order of the records correspond to their logical order and that reuses the space which had been occupied by records which were deleted.

When a file is treated as a succession of blocks, efficiency depends on the blocks staying about equal in size. This is very simple to manage if the key is alphabetic and the number of records with keys that begin with the same letter stays about the same for each of the letters A to Z. Then the file can be divided into 26 blocks, one for each letter of the alphabet. The allocation of a record to a block is made just by reference to the first letter of its key. On average, deletions and additions would balance out. Unfortunately, matters are seldom so simple and other criteria are used to decide, from the value of the key, the block into which a record should be put. The criteria are embodied in what are called *hashing algorithms*.

These considerations can be ignored by using yet another programming resource that is illustrated by the "random" files of IBM PC BASIC. You just assign identification numbers to records when they are stored, or you let the system assign consecutive numbers by default. The system maintains a directory that relates the identification numbers to the physical locations of the records. To retrieve a record, you simply give its identification number. Thus, in IBM PC BASIC, you can (1) open a file for random access, (2) use LSET and RSET statements to set up a record that will be transferred to that file, and (3) transfer the record by a statement of the form PUT #n,c%, where n is the file number and c% is the record number you want the record to have. The record is retrieved by the corresponding statement of the form GET #n,c%. When you store the updated version of the record by a PUT statement that contains the same identification number, you do not need to know how the record was stored on disk. From the standpoint of the user, it does not even matter whether the records with successive record numbers are physically adjacent.

Putting records into storage wherever space happens to be available simplifies the programs that maintain the file. There is, however, an attendant overhead in the coding that relates record number to physical location. When there is a frequent need to process all the records in sequence on one particular key, the relative simplicity of processing records in physically sequential order, which prevails with indexed sequential files, is absent.

There is a vast amount of further detail concerning these issues of storage control. The bibliography provides some suitable references.

The other major topic considered in this chapter deals with merging and related operations. The principle of merging, described in Section 2.3, can be extended very easily to records that contain several fields. The files that are merged must be in order on the same key. The first record of each file is read, and the keys are compared. Depending on which key has precedence, one record or the other is written to the output file. Merging may combine two files of comparable size that have come from different sources. Thus, you may wish to merge personnel files of several divisions of a company or the holdings files of several libraries that are pooling their resources.

The commonest way to incorporate new records into a file also is to merge, thereby combining an old file that may be quite lengthy with a sorted batch of additions that is relatively short. This process can be extended further to delete and to change specified records in the old file. Thus, each record in the updated file may be either (1) a complete record, showing that the program is to insert it in the new file and omit the corresponding record of the old file if one with the same value of the key is present, (2) a change record that begins with a special code and contains a specification of the field(s) to be changed and the new value(s), or (3) a deletion record that consists of a code that signifies its role and the value of the key in the record that is to be deleted.

An important sorting technique also is based on merging. The simplest form of this "merge sort" can be illustrated with a file of 1024 elements as follows. Call this file A. To start, it is split in the middle into two files B and C. These are combined into a new file A in which the first element of B and the first element of C are in order with respect to each other; so are the second element of B and the second element of C, and so on to the five hundred twelfth elements of B and C. The new file A is split in the middle into two new files B and C. These are combined into a new file A by merging the first and second elements of B with the first and second elements of C, merging the third and fourth elements of B with the third and fourth elements of C, and so on to the two hundred fifty-fifth and fifty-sixth elements of the two files. File A is split in the middle once more, into two new files B and C. This time the elements are merged in groups of four. The process of splitting and merging is repeated six more times, thereby merging groups of 8, 16, 32, 64, 128, 256, and 512 elements, respectively. The final file is in order. This principle is used in several ways that are more elaborate and faster. For example, by splitting the file in four at the start of each cycle, only four cycles are needed. These bring groups of 4, 16, 64, and 256 elements together in order.

This use of merging in an external sort algorithm shows that the methods of the present chapter can be applied to the problems of Chapter 1. The sorting module in the program of Section 2.3 illustrates the bearing of the earlier material on the topics we have been discussing. The storage allocation problems discussed earlier in this section are a major reason for the use of the linked lists with which Chapter 3 is concerned.

Suggested Activities

1. Design and store a file of reference data on disk. Sort it on an appropriate field, and construct an index file that consists of the values of the key in the successive records. Write a program that (a) loads the index from disk, (b) prompts for a value of the key, (c) searches the index for the key, (d) retrieves the appropriate record without making any more comparisons, and (e) repeats steps (b) to (d) until instructed to quit. Run some experiments to compare the time taken by this method with the time taken by an external serial search. Expand the program to batch several search requests, sort them in sequence, and retrieve the required records in a single pass through the main file.

2. Design and store a body of presorted reference data on disk in a succession of short files. Construct an index to the first key value in each of the files. Write a program to use the index to deal with a search request by (a) searching the index to find the appropriate file and (b) searching the file.

3. Expand Activity 2 to include detailed indexes that point from other keys to the key on which the data is sequenced, and which search on those other keys.

4. Expand Activity 2 to allow new records to be added to a supplementary file and periodically sorted and merged into the main files.

5. Write a program to perform the merge sort (a) by using a two-way split and merge in each cycle and (b) by using a four-way split and merge.

3

Working with Linked Lists

The physical order of the items in a list has been a major concern throughout Chapters 1 and 2. The programs in Chapter 1 sort the contents of an array by moving them around until the items with successive values of the subscript are in numerical or alphabetic sequence. The programs to search, merge, and perform other processes in Chapter 2 use the fact that items that came in physical sequence are also in sequence numerically or alphabetically.

Maintaining that kind of physical order can be difficult, however, in a list or a file that undergoes frequent additions and deletions. This problem was mentioned in Section 2.4. The present chapter describes the linked list technique, which avoids the need to keep a file in a physical order that parallels its alphabetic or numerical order. Instead, each item is accompanied by the serial number of the item that comes next numerically or alphabetically. We call this number a link or pointer and comment on the terminology at the end of the chapter. Its inclusion lets you work rapidly through the items in numerical or alphabetic sequence when they are stored in an array. When each item is a record that consists of several fields, moreover, you can include separate links that correspond to the order of the data in those fields.

Linked lists play a major role in many computer applications. The basic concept has been extended in several ways. Special features have been included in several programming languages and in the actual design of some computers, to facilitate the use of linked lists.

The programs in this chapter let users proceed step by step through the basic linked list operations. An extensive set of prompts explains what is about to happen at each stage. A tie line is drawn from one element of a list to its logical successor when the focus of attention changes. The colors of individual items are also changed to reflect their changing roles in the overall process. These display techniques, and the underlying programming methods, can be used in many other instructional and related applications, too.

3.1   Building a Linked List

The program LINKADD provides a simple introduction to the linked list principle. LINKADD lets you type a series of names in any sequence but keeps track of the alphabetic order of the names by means of a set of links.

The program begins by displaying a message that it is setting up some data. After a brief pause the display changes to

```
List is empty            1 ----   /

                         2 ----   /

                           . . .

Please type:            10 ----   /

A to add a name,        11 ----   /

T to trace a list,      12 ----   /

Q to quit:
```

The right-hand side of the display depicts the array in which the names will be stored. The numbers 1 to 12 identify the successive elements of the array. These are called slots in the prompts. The hyphens show that the slots are empty. The slashes depict links that have not been set yet and which will serve a role that is the central topic of this chapter.

You can make a simple and rapid start by entering a few names in reverse alphabetic order. Press A to add a name, and the prompt in the lower-left corner of the screen changes to tell you to type a name (the limit is four letters). Type Zeno. You are prompted to press the space bar to store the new name in slot 1 of the array that holds the linked list. The line for slot 1 changes to:

```
1 Zeno  /
```

You are prompted to press the space bar to make the new entry into the new tail. The line changes to

```
1 Zeno  0
```

The 0 shows that Zeno is the tail; it is the alphabetically final element of the list at this time. Since Zeno is the only name in the list, making this determination has not taxed the program excessively. The message changes to "press the space bar to continue." You respond, and the display changes to

```
Head of list is       h→  1  Zeno   0

  in slot 1                2  ----   /

Tail of list is            3  ----   /

  in slot 1                4  ----   /

                            . . .

Please type:              10  ----   /

A to add a name,          11  ----   /

T to trace a list,        12  ----   /

Q to quit:
```

The head of the list is simply the name that comes first in it alphabetically. Because the list contains just one element at this time, the head and the tail are the same. Type A, and the prompt for a name reappears at the lower-left corner. Enter the name Will. The program prompts you to press the space bar to store the new name in slot 2.

The program LINKADD always puts a new name into the first available slot. Then it sets the relevant links to show the alphabetic position of this name relative to the other names in the list. To see this, press the space bar. The top two lines change to:

```
                 1  Zeno   0

                 2  Will   /
```

The new name is displayed in white on black, shown here by boldface. You are prompted to press the space bar to start searching. The program is about to embark on a systematic search through the names that are already in the list. This will find where the new name fits alphabetically. You press the space bar. The head name, that is, Zeno, changes to purple on black (shown here by underlining). This shows that it has been selected for comparison with the new name. You are prompted to press the space bar to compare the new name with the highlighted entry.

You press the space bar and the computer compares the new name, Will, with the name just selected, Zeno. The program recognizes that Will precedes Zeno alphabetically, and the prompt changes to press the space bar to link the new name to the highlighted entry. You respond.

The pointer alongside Will is set to 1 to show that Will is followed alphabetically by the contents of slot 1. A tie line is drawn to reinforce this fact.

```
  ┌─> 1  Zeno   0
  │   ─         ─
  └───────────────
      2  Will   1─┘
```

You are prompted to press the space bar to continue. The tie line from Will to Zeno disappears as soon as you respond. The current state of the list appears:

```
Head of list is           1 Zeno   0

in slot 2           h─→   2 Will   1

Tail of list is           3 ----   /

in slot 1                 4 ----   /
```

The prompt to type A, T, or Q reappears; type A to add another name—Vera. Pressing the space bar, in response to successive prompts:

1. Puts Vera into slot 3
2. Compares Vera and Will alphabetically
3. Finds that Vera precedes Will
4. Sets the link alongside Vera to 2 to point to the slot that Will occupies
5. Draws the tie line from Vera to Will

```
      1  Zeno   0

  ┌─> 2  Will   1
  │   ─         ─
  └───────────────
      3  Vera   2─┘
```

The wording of the prompts explains what is happening, and color changes in the depiction of the list focus attention on the relevant entries. The link from Will is still 1, which shows that the entry in slot 1, that is, Zeno, follows Will alphabetically.

You can continue in this way: typing A in response to the main prompt and entering the names Ulf, Tom, Sal, Ron, Rob, Rab, Pete, Paul, and Pat in succession. The program repeats the pattern of activity that has been described. In each cycle, the search goes no further than the former head of the list, because the new name precedes it alphabetically. After the name Pat has been processed, the list is displayed as follows:

```
Head of list is          1  Zeno   0

in slot 12               2  Will   1

                         3  Vera   2

Tail of list is          4  Ulf    3

in slot 1                5  Tom    4

                             . . .

                     h→12  Pat    11
```

If you try to enter another name, the program displays a message
that the list is full and that pressing the space bar will bring back the
main menu. You can type T to trace the list, with the effect described
later, or you can type Q to quit. Then you can restart the program and
continue taking it through its paces in a systematic fashion. Thus, you
can enter a series of names in alphabetic order. As an example, start
with Abe. This shows nothing new. Now enter the name Al. This time,
the comparison of the new name with the first (and only) name already
in the list triggers the prompt that tells you the new name is alphabetically higher. Then the program finds it has reached the end of the
names already in the list. This triggers the prompt to make the new
entry into the new tail. A tie line is drawn from Abe to Al. The link on
the right of Abe changes from 0 to 2 to show that Abe is followed by the
name in slot 2. The link on the right of Al is set to 0, which shows that
Al is the new tail.

```
        1  Abe    2─┐
        ─           ─ ┘
        └─> 2  Al     0
```

When you press the space bar to continue, the tie line disappears. The
message that reappears at the top left of the screen also shows that the
tail is now in slot 2. Now enter the name Amy. This goes into slot 3.
The comparison of Amy and Abe shows that the new name is alphabetically higher. Consequently, you are prompted to find the next entry. A
tie line is drawn from Abe to the alphabetically next name, Al.

```
        1  Abe    2─┐
        ─           ─ ┘
        └─> 2  Al     0

            3  Amy    /
```

The name that follows Abe alphabetically is in the line that follows it
on the screen, because we entered names in alphabetic order in this

example. The prompt to make a comparison reappears. The program
finds that the new name again follows the name it has been compared
with alphabetically. The tie line from Abe to Al disappears, and then
the program finds it has reached the end of the names already in the
list. It displays the prompt to make the new name into the new tail.
The program draws the tie line from Al to Amy, changes the link on
the right of Al from 0 to 3, sets the link on the right of Amy to 0, and
changes the slot number of the tail to 3. In these steps, color changes
mark the shift of attention from name to name.

```
Head of list is        h↦ 1 Abe    2

in slot 1                 2 Al     3

Tail of list is           3 Amy    0

in slot 3                 4 ----   /
```

This display shows that the list starts alphabetically with the name in
slot 1, which is followed alphabetically by the name in slot 2, which is
followed alphabetically by the name in slot 3, which ends the list.

Now enter the name Anne, which goes into slot 4. The head entry
Abe is compared with the new name, and the new name is found to be
higher alphabetically. The search moves on to Al. The new name Anne
is again found to be higher alphabetically. It is compared with Amy.
Once again the comparison shows that the new entry is higher. Now
the end of the existing list has been reached. Amy is linked to Anne,
which becomes the new tail, in much the same way that the tail was
changed previously.

Entering further names in alphabetic order follows the pattern that
has been described, but with an increasing number of steps in each
search. Whereas successive names required just one comparison when
they went to the head of the list, names that go to the tail must be
compared in turn with every name that is in the list already. If you
keep going in alphabetic order, entering the names Bart, Ben, Bert,
Bill, Bob, Carl, Don, and Ed, the available space fills and the display
becomes:

```
Head of list is        h↦ 1 Abe    2

in slot 1                 2 Al     3

Tail of list is           3 Amy    4

in slot 12                4 Anne   5

                           . . .

                          12 Ed    0
```

Now you can quit, restart, and proceed to demonstrate the generality
of the process. Enter the name Hans, and then enter the name Mary,

pressing the space bar enough times to set the links for these two items. Then enter the name Jane. The first three lines now are:

```
1  Hans   2

2  Mary   0

3  Jane   /
```

Compare Jane with Hans. Jane is higher. Because the link from Hans is 2, the program goes on to compare Jane with Mary. When you press the space bar again, the program finds that the new entry Jane precedes Mary alphabetically. The program prompts for a key press to link Jane to Mary, sets the link from Jane to 2, and draws the tie line.

```
      1  Hans   /

  ┌─> 2  Mary   0
  └─────────────
      3  Jane   2─┘
```

The program has "remembered," however, that the new entry also follows Hans alphabetically. You are prompted to link the former predecessor of Mary—Hans—to the new entry. Pressing the space bar makes the program erase the tie line from Jane to Mary, draw the tie line from Hans to Jane, and change the pointer from Hans to 3.

```
      1  Hans   3─┐
                  │
      2  Mary   0 │
  ┌───────────────
  └─> 3  Jane   2
```

These examples illustrate the three basic situations that can occur when a new entry is inserted into a linked list. It can become the new head, it can become the new tail, or it can come between two entries already in the list. The positions of the head and the tail, and the relative positions of each entry and its alphabetic predecessor and successor, are quite unrestricted. For example, if you enter the names Hans, Gary, Pete, Al, Fred, and Ron in that order, the partial lists when the successive names have been typed are as follows.

```
h→  1 Hans  0         1 Hans  0           1 Hans  3
                  h→  2 Gary  1       h→  2 Gary  1
                                          3 Pete  0


    1 Hans  3         1 Hans  3           1 Hans  3
    2 Gary  1         2 Gary  1           2 Gary  1
    3 Pete  0         3 Pete  0           3 Pete  6
h→  4 Al    2     h→  4 Al    5       h→  4 Al    5
                      5 Fred  2           5 Fred  2
                                          6 Ron   0
```

Typing T in response to the main menu lets you trace through a list entry by entry starting at the head. Successive entries are (1) emphasized by a color change, (2) joined to the alphabetically next entry by a tie line, and (3) deemphasized when you press the space bar after successive prompts.

Program Variables

HCP	The horizontal cursor position of the 10s digit of the slot numbers
S$	A string of 40 spaces used in PRINT statements to force double spacing
RA$	The right arrow graphics character
KLIM	The number of different prompting messages displayed at the lower-left corner of the screen
M$(k,0)	The number of lines of text in the kth message
M$(k,j)	The jth line of the kth message, when j>0
K	The numbers 1 to KLIM for the different messages, respectively
J	The numbers 1, 2, . . . for successive lines of text within a message
FN P$(N)	The character string that (1) consists of a slash when P(N) is negative and represents N when N is nonnegative and (2) is right-aligned in a two-character field in both cases
NLIM	The maximum length of the lists handled by the program (the limit is set by the screen display, not by the underlying logic)
A$(n)	The name occupying the nth slot in the array that stores the names that occur in the list
P(n)	The link from slot n to the slot that contains the successor of A$(n) in the alphabetic ordering of items in the list, when n>0, and the slot number of the head of the list when n is 0
NF	The number of the slot where the next item that is added to the list will be stored
NS	The slot number of the entry just added to the list
TAIL	The number of the slot that contains the final item of the list in alphabetic order
R$	The one-byte response to certain prompts
OP	The numerical operation code that is 1, 2, or 3 to add, trace, or quit, respectively
UNPLACED	The switch that is 1 and 0, respectively, before and after the position of the new item has been found

M	The number of the slot from which a tie line is drawn
NN	The dummy slot number in a subroutine that displays the contents of a slot
LINK$	The string of graphic characters that represents the tie line
Q	The point where the LINK$ string must be split to prevent a spurious displacement when it is displayed on the screen
CR$,CL$	The "characters" that move the cursor one pointer to the left and right, respectively, when encountered in the operand of a PRINT statement.
CU$,CD$	The "characters" that move the cursor up and down, respectively
SW$,NW$	The graphics characters that form the southwest and northwest quarters of a cross that spans a character position
SE$,NE$	The graphics characters that form the southeast and northeast corners of such a cross
VB$,HB$	The graphics characters that appear as vertical and horizontal lines through the center of a character position
DNS$,UPS$	The strings of graphics characters that form the first part of the tie lines extending downward and upward, respectively
DNE$,UPE$	The corresponding strings for the second part of the down and up tie lines, respectively
DCS$,UCS$	The strings of spaces and cursor control codes used to erase the first part of down and up tie lines
DCE$,UCE$	The corresponding strings used to erase the second part of these tie lines
I	The index of some loops that refer to successive lines of the display

Program Notes

The program consists of:

1. A fairly substantial initialization section
2. The main action cycle, which embodies the algorithmic process of inserting names into a linked list by using (3)
3. The subroutine that works through the list, entry by entry
4. A collection of subroutines to deal with different parts of the display
5. The error trap and quit sections

The initialization section performs some general housekeeping, sets P(0) to 0 to signify that the list is empty, sets NF to 1 to specify that the first name to be entered will go into slot 1, sets TAIL to 0, consistent with the list being empty, and displays the empty list.

The main part of the program displays data about the current head and tail, displays the main menu, and sets OP in accordance with the key that is pressed. When this is A, to add a new name, the slot number NS of the new entry is set to NF. Then NF is increased by 1, to serve as slot number of the next entry, or is reset to 0 if the list is full.

```
2100  NS=NF
. . .
2130   IF NF<NLIM THEN NF=NF+1 ELSE NF=0
```

The search subroutine is executed to find the alphabetic position of the new entry.

```
2400 GOSUB 5000
```

Within the subroutine, a WHILE . . . WEND loop controls the search. The loop works through the list starting at the head entry. The program initializes P(0) to 0 to signify an empty list. Thereafter, it keeps the slot number of the head in P(0). The statements that take care of the actual searching are as follows:

```
5010   UNPLACED=-1
5020   N=0
5030   WHILE UNPLACED AND P(N)>0
. . .
5170     IF OP=1 THEN UNPLACED = (A$(P(N))<A$)
5180     IF NOT UNPLACED THEN 5260
. . .
5250     N=P(N)
5260   WEND
```

These work in the following manner. Suppose first that the list is empty. The loop is bypassed, leaving both N and P set to 0. Suppose next that the list contains at least one element and that the new item precedes the old head of the list alphabetically or is the same. As a result, UNPLACED is reset to 0 in the first cycle through the loop. Consequently, the computer goes through the loop just once and leaves with N still set to 0.

Suppose next that the new item follows the first item in the list alphabetically and precedes the second item (or is equal to it). The first cycle through the loop causes N to be reset to P(0), that is, to the slot number of the head. The second cycle through the loop resets UNPLACED to 0. The loop is left with N set to the slot number of the head. Thus, N points to the predecessor of the item that is alphabetically higher (or equal to) the new item.

Now suppose that the new item follows the item in slot N' alphabetically and precedes (or is equal to) the next item already in the list. Successive cycles of the loop lead to the circumstance that N equals N'. In consequence, P(N) points to the item that makes the comparison reset UNPLACED to 0. The loop is left, accordingly, with N pointing to the predecessor of the entry that must follow the new name.

Finally, suppose that the new item is alphabetically higher than the entry at the end of the list. Successive cycles through the loop lead to the circumstance that N is the slot number of the predecessor of the tail. The test A$(P(N))<A$ fails again. N is reset to point to the tail. The WEND sends control back to the WHILE statement. This time, because N points to the tail, P(N) is zero. Control jumps past the WEND. The further statements in the WHILE ... WEND loop provide the visual effects in all these cases.

Control thus reaches the "reset-links" section (1) with N = 0 and P(N) = 0, when the list was empty, (2) with N = 0 and P(N)>0, when the new item precedes the old head alphabetically, or is the same, (3) with N>0 and P(N)>0, when the new item fits alphabetically between the items in slots N and P(N), or is the same as the item in slot P(N), or (4) with N>0 and P(N) = 0, when the new item follows the former tail alphabetically. The reset-links section sets the link from the new item, that is, P(NS), to the former value of the link P(N). In cases 1 and 4, setting P(NS) marks the new item as the new tail, because the former value of P(N) was 0. In cases 2 and 3, setting P(NS) links the new item to the former head, or to a later element in the list, respectively. In cases 1 and 4, also, TAIL is reset to the slot number of the new item. In all four cases, P(N) is reset to point to the new item. In cases 1 and 2, setting P(N) designates the new item as the head, because P(0) serves as the pointer to the head and N is 0.

These actions are performed very simply by the statements:

```
2550   P (NS) =P (N)
2560   IF  P (N) =0  THEN  TAIL=NS
2680   P (N) =NS
```

The search subroutine is also used to trace through the successive elements of a list. The test to reset UNPLACED is applied only when OP is 1.

```
5170   IF  OP=1  THEN  UNPLACED  ...
```

When tracing a list, OP is 2 and UNPLACED is not reset. Consequently, the WHILE . . . WEND loop then works through the entire list. Thus, a very small number of statements takes care of the algorithmic process, though these do require a certain amount of explanation.

Suggested Activities

1. Adapt LINKADD to display an expanded main menu and, when requested, to display:
   (a). The length of the list
   (b). The alphabetic position of a name that you type in response to a further prompt (or 0 if the name is not in the list)
   (c). The number of occurrences of a given name in the list
   (d). The successor of a given name
   (e). The predecessor of a given name
   (f). The name that occurs in the nth position alphabetically (after you have typed the value of n in response to a further prompt)
   (g). The names in the list that fall alphabetically between two names that you type in response to a further prompt
   (h). The number of items that follow a given name in the list
   In these adaptations omit the visual display and the step-by-step cycling if you wish, increase the size of the array, and include the capability to display the elements of the list in alphabetic order.

2. Adapt LINKADD:
   (a). To ignore the request to add a name to the list if the name is in the list already
   (b). To merge two linked lists
   (c). To display the elements of a linked list in reverse order

3. In a doubly linked list, each item is accompanied by a link to its predecessor as well as to its successor. In a circular linked list, the tail is linked to the head. Expand LINKADD to demonstrate the insertion of elements into (a) a doubly linked list, (b) a circular linked list, and (c) a circular doubly linked list.

4. Expand the program(s) written in accordance with Activity 3 to perform the actions specified in Activities 1(a) to (g) and 2(a) to (d).

5. Adapt LINKADD to maintain two linked lists within the same array and to permit the insertion of a name into list 1 or list 2 and the transfer of a name from list 1 to list 2, or vice versa. (Hint: Use separate variables for the slot numbers of the heads of lists 1 and 2.)

```
100 REM * LINKADD inserts into a linked list without deletion *
101 :
110  ON ERROR GOTO 10000                                      'error trap
120   KEY OFF: SCREEN Ø
130   WIDTH 4Ø: COLOR 7,1,1                                   'screen set up
140   CLS
150  LOCATE 1Ø,8,Ø,Ø
160   PRINT "Linked list demonstration"
170  LOCATE 14,8,Ø,Ø
180   PRINT "Please wait - setting up data"
200   HCP=24                                          'position of list
210   S$=SPACE$(4Ø)
220   RA$=CHR$(26)                                            'right arrow
230   GOSUB 9ØØØ                                      'strings for tie lines
299 :
300 REM * set up the message arrays *
310   KLIM=21                                                 '# of messages
320   DIM M$(KLIM,8)                                          'message array
330  FOR K=1 TO KLIM
340   READ M$(K,Ø)                                            'message length
350   FOR J=1 TO VAL(M$(K,Ø))                                 'transfer lines
360    READ M$(K,J)                                           'from data stmts
370  NEXT J,K
499 :
500 REM * set up the array *
510   CLS
520   DEF FN P$(N)=MID$(" /"+RIGHT$(STR$(P(N)),2),3+2*(P(N)<Ø),2)   'link or /
530   NLIM=12                                                 'capacity
540   DIM A$(NLIM),P(NLIM)                            'entries & pointers
550  FOR N=1 TO NLIM
560   A$(N)="----"                                            'null entry
570   P(N)=-1                                                 'link
580   GOSUB 6Ø1Ø                                      'display them
590  NEXT N
600   NF=1                                            'pointer to free space
610   TAIL=Ø                                                  'tail of list
620   P(Ø)=Ø                                                  'head of list
999 :
1000 REM * branch for action *
1010   FOR J=1 TO 1Ø
1020    LOCATE J,1,Ø,Ø                                 'clear the prompt area
1030     PRINT SPC(HCP-3)
1040   NEXT J
1050  LOCATE 1,1,Ø,Ø                                          'specify the head
1060   IF P(Ø)=Ø THEN PRINT "List is empty"
                 ELSE PRINT "Head of list is" S$ "in slot" P(Ø)
1070   PRINT
1080   IF TAIL>Ø THEN PRINT "Tail of list is" S$ "in slot" TAIL       'and tail
1100   IF P(Ø)>Ø THEN LOCATE 2*P(Ø)-1,HCP-2: PRINT "h" RA$
1120   K=1: GOSUB 65ØØ                                  'prompt for action
1130    R$=INPUT$(1)
1140   OP=INT((INSTR("AaTtQq",R$)+1)/2)                       'op code
1150    IF OP=Ø THEN OP=1                                      'default
1160   LOCATE 1,1,Ø,7
1170   ON OP GOTO 2ØØØ,4ØØØ,12ØØØ                             'branch
1999 :
2000 REM * insert a new item into the list *
2010   IF NF>Ø THEN 2Ø5Ø
2020    K=2Ø: GOSUB 65ØØ                               'no room for another item
2030    Z$=INPUT$(1)
```

```
2040    GOTO 1000
2050    K=2: GOSUB 6500                              'prompt for the item
2060    LOCATE CSRLIN,POS(0),0,7                         'to be inserted
2070    INPUT; " ",A$
2080    K=3: GOSUB 6500                      'prompt for key press to store it
2090    PRINT NF;
2100    NS=NF                                        'set the slot #
2110    Z$=INPUT$(1)
2120    A$(NS)=A$                                    'store the new item
2130    IF NF<NLIM THEN NF=NF+1 ELSE NF=0    'reset slot for next item
2180    IF P(0)=0 THEN LOCATE 1,6: PRINT "was empty": GOTO 2230
2190    LOCATE 1,14                                      'alter the
2200     PRINT "was"                         'comment about the head
2210    LOCATE 2*P(0)-1,HCP-2                    'erase its marker
2220     PRINT " "
2230    P(NS)=-1                                    'undefine the link
2240    COLOR 7,0
2250    GOSUB 6020                              'highlight the new item
2270    IF TAIL=0 THEN 2400
2280    LOCATE 5,14
2290     PRINT "was"                          'comment about the tail
2400    GOSUB 5000                              'search the linked list
2499 :
2500 REM * reset the links *
2510    IF P(N)=0 THEN 2550                 'jump if new item is new tail
2520    K=10: GOSUB 6500
2530    Z$=INPUT$(1)                        'prompt to link predecessor of
2540    GOSUB 7500                          'item in slot N to new item
2550    P(NS)=P(N)                             'set link from new item
2560    IF P(0)=0 THEN TAIL=NS: GOTO 2610     'new item is new tail
2570    M=NS: GOSUB 7000                    'draw tie line from new item
2580    COLOR 7,0                               'de-highlight the item
2590    GOSUB 6020
2600    IF N=0 THEN 2680                    'jump if new item is new head
2610    IF P(N)>0 THEN K=11 ELSE K=12: GOTO 2650   'prompt to link to new item
2620    P(N)=-1                             'undefine link in slot N
2630     COLOR 3,0                          'redisplay contents of slot N
2640     GOSUB 6010
2650     GOSUB 6500                             'display the prompt
2660    Z$=INPUT$(1)
2670     GOSUB 7500                             'erase the tie line
2680    P(N)=NS                             'reset link to new item
2690    GOSUB 6020                             'redisplay the item
2700    M=N: GOSUB 7000                        'display the tie line
2710    IF N>0 THEN GOSUB 6010             'de-highlight the linked item
2720    IF P(NS)>0 THEN NN=P(NS): GOSUB 6200    'de-highlight the successor
2730    K=13: GOSUB 6500                       'prompt to continue
2740    Z$=INPUT$(1)
2750    GOSUB 7500                             'erase the tie lines
2760    M=NS
2770    GOSUB 7500
2780    GOTO 1000
3999 :
4000 REM * trace the list *
4010    IF P(0)>0 THEN LOCATE 2*P(0)-1,HCP-2: PRINT " "        'erase h and
4020    IF NF>0 THEN LOCATE 2*NF-1,HCP-2: PRINT " "            'f marker
4030    GOSUB 5000                             'work through the list
4040    GOSUB 6010                         'de-highlight final entry
4050    GOTO 1000
4999 :
```

```
5000 REM * search the list *
5010   UNPLACED=-1                                        'set the found switch
5020   N=0                                                'start at head of list
5030   WHILE UNPLACED AND P(N)>0                          'loop through elements
5040     IF N=0 THEN K=4 ELSE K=5
5050       GOSUB 6500                                     'prompt for next action
5060       Z$=INPUT$(1)
5070     IF N=0 THEN 5120                                 'jump if first element
5080       M=N
5090       GOSUB 7000                         'draw tie line from predecessor
5100       COLOR 3,0
5110       GOSUB 6010                                 'de-highlight predecessor
5120       COLOR 13,0
5130       GOSUB 6030                                       'highlight successor
5140     IF OP=1 THEN K=6 ELSE 5190                        'prompt for next action
5150       GOSUB 6500
5160       Z$=INPUT$(1)
5170     IF OP=1 THEN UNPLACED = (A$(P(N))<A$)                          'compare
5180     IF NOT UNPLACED THEN 5260                            'jump if match
5190     IF OP=1 THEN K=6 ELSE K=13                      'prompt for next action
5200       GOSUB 6500
5210       Z$=INPUT$(1)
5220     IF N=0 THEN 5250                                 'jump if 1st element
5230       GOSUB 6010                                           'redisplay
5240       GOSUB 7500                                           'erase tie line
5250       N=P(N)                                      'pointer to next element
5260   WEND
5270   RETURN
5999 :
6000 REM * display lines of list
6010   NN=N:    GOSUB 6200: RETURN                        'contents of slot N
6020   NN=NS:   GOSUB 6200: RETURN                        'slot for new item
6030   NN=P(N): GOSUB 6200: RETURN                        'slot for successor
6200 REM * display row nn *
6210   IF NN<1 THEN RETURN                                'bypass if fictional
6220   LOCATE 2*NN-1,HCP,0,0
6230     PRINT USING "## \  \ \\"; NN,A$(NN),FN P$(NN)              'display
6240   COLOR 7,1                                            'default colors
6250   RETURN
6499 :
6500 REM * display a message *
6510   FOR I=15 TO 25 STEP 2
6520     LOCATE I,1,0,0                                   'clear message area
6530     PRINT SPC(HCP-5)
6540   NEXT I
6550   FOR J= 1 TO VAL(M$(K,0))                    'loop through lines of prompt
6560     LOCATE 25-2*(VAL(M$(K,0))-J),1,0,0
6570     PRINT M$(K,J);
6580   NEXT J
6590   RETURN
6999 :
7000 REM * draw the connecting line *
7010   IF M<=0 OR P(M)<=0 THEN RETURN          'bypass if either end fictional
7020   LOCATE 2*M-1,HCP+10,0,0                            'starting position
7030   IF M<P(M) THEN LINK$=LEFT$(DNS$,6*(P(M)-M-1)+5)+DNE$
                  ELSE LINK$=LEFT$(UPS$,6*(M-P(M)-1)+5)+UPE$    'tie line string
7040   WHILE LEN(LINK$)
7050     Q=40-POS(0)+1                                    'break point
7060     PRINT LEFT$(LINK$,Q);                            'display a section
7070     LINK$=MID$(LINK$,Q+1)                            'residual portion
```

```
7080    WEND
7090    RETURN
7500 REM * erase the connecting line *
7510    IF M<=Ø OR P(M)<=Ø THEN RETURN
7520    LOCATE 2*M-1,HCP+1Ø,Ø,Ø
7530    IF M<P(M) THEN LINK$=LEFT$(DCS$,6*(P(M)-M-1)+5)+DCE$
                 ELSE LINK$=LEFT$(UCS$,6*(M-P(M)-1)+5)+UCE$    'blank out string
7540      WHILE LEN(LINK$)
7550        Q=4Ø-POS(Ø)+1                                    'break position
7560        PRINT LEFT$(LINK$,Q);                            'display a section
7570        LINK$=MID$(LINK$,Q+1)                            'residual portion
7580      WEND
7590    RETURN
7999 :
8000 REM * actual messages *
8010    DATA 4, "Please type:", "A to add a name,", "T to trace a list,",
                "Q to quit:"                                            :'1
8020    DATA 3, "Type a name",    "(limit is 4", "letters):"           :'2
8030    DATA 4, "Press the space", "bar to store the", "new name in", "slot"  :'3
8040    DATA 3, "Press the space", "bar to start", "searching"         :'4
8050    DATA 3, "Press the space", "bar to find the", "next entry"     :'5
8060    DATA 5, "Press the space", "bar to compare", "the new name with",
                "the highlighted", "entry"                             :'6
8070    DATA 1, "dummy message for consistency in numbering"           :'7
8080    DATA 1, "dummy message for consistency in numbering"           :'8
8090    DATA 1, "dummy message for consistency in numbering"           :'9
8100    DATA 4, "Press the space", "bar to link the", "new entry to the",
                "highlighted entry"                                    :'1Ø
8110    DATA 6, "Press the space", "bar to link the", "predecessor of",
                "the highlighted", "entry to the", "new entry"         :'11
8120    DATA 4, "Press the space", "bar to make the", "new entry into",
                "the new tail"                                         :'12
8130    DATA 2, "Press the space", "bar to continue"                   :'13
8140    DATA 1, "dummy message for consistency in numbering"           :'14
8150    DATA 1, "dummy message for consistency in numbering"           :'15
8160    DATA 1, "dummy message for consistency in numbering"           :'16
8170    DATA 1, "dummy message for consistency in numbering"           :'17
8180    DATA 1, "dummy message for consistency in numbering"           :'18
8190    DATA 1, "dummy message for consistency in numbering"           :'19
8200    DATA 3, "List is full.", "Press space bar", "for main menu."   :'2Ø
8210    DATA 3, "List is empty.", "Press space bar", "for main menu."  :'21
8999 :
9000 REM * strings used to draw link lines *
9010    CR$=CHR$(28) : CL$=CHR$(29)                          'cursor right & left
9020    CU$=CHR$(3Ø) : CD$=CHR$(31)                                   'up & down
9030    SW$=CHR$(191): NE$=CHR$(192)                    'right angle characters
9040    NW$=CHR$(217): SE$=CHR$(218)
9050    VB$=CHR$(179): HB$=CHR$(196)               'vertical & horizontal bars
9060    DNS$=HB$+SW$                              'start 1st part of tie lines
9070    UPS$=HB$+NW$
9080    UCS$="  "                                       'start of erasure lines
9090    DCS$="  "
9100    FOR I=1 TO 11
9110      DNS$=DNS$+CL$+CD$+VB$+CL$+CD$+VB$                     'continuations
9120      UPS$=UPS$+CL$+CU$+VB$+CL$+CU$+VB$
9130      UCS$=UCS$+CL$+CU$+" "+CL$+CU$+" "
9140      DCS$=DCS$+CL$+CD$+" "+CL$+CD$+" "
9150    NEXT I
9160    DNE$=CL$+NW$                               'start 2nd part of tie lines
9170    UPE$=CL$+SW$
```

```
9180   UCE$=CL$+" "                                              'and erasure lines
9190   DCE$=CL$+" "
9200    FOR I=1 TO 14
9210      DNE$=DNE$+CL$+CL$+HB$                                  'continuations
9220      UPE$=UPE$+CL$+CL$+HB$
9230      DCE$=DCE$+CL$+CL$+" "
·9240     UCE$=UCE$+CL$+CL$+" "
9250   NEXT I
9260   DNE$=DNE$+CL$+SE$+CL$+CD$+NE$+HB$+">"                     'arrow heads
9270   UPE$=UPE$+CL$+NE$+CL$+CU$+SE$+HB$+">"
9280   DCE$=DCE$+CL$+" "+CL$+CD$+"    "
9290   UCE$=UCE$+CL$+" "+CL$+CU$+"    "
9300    RETURN
9999 :
10000 REM * error trap *
10010   LOCATE 25,1
10020    PRINT "Error" ERR "in line" ERL "- press space bar";   'diagnostic
10030    Z$=INPUT$(1)
10040   RESUME 12000
11999 :
12000 REM * quit *
12010   SCREEN 0: WIDTH 80: COLOR 7,0,0                          'default screen
12040    CLS
12050   LOCATE 1,1,1,7                                           'restore cursor
12060   ON ERROR GOTO 0                                          'reset error address
12070   END
```

6. Adapt LINKADD to maintain n linked lists in the same array.
   When a new name is added, specify the list into which it should be
   inserted.

7. Write a program that stores the titles and authors of a collection of
   books in an array. Maintain links that show the alphabetic order of
   (*a*) the books and (*b*) the authors.

8. Use a linked list approach to sort an array of items without moving
   them.

3.2 Linking the Space That Is Free

The program LINKDEL demonstrates the second component of the
general process of maintaining a linked list. The program LINKADD
in the preceding section showed how to keep track of the alphabetic
ordering of items as they were added to a list. The problem of deleting
items is demonstrated by LINKDEL, which, however, keeps track only
of the order of arrival. The program LINKLIST in the next section
shows how to add, keep track of alphabetic order, and delete items in a
linked list by combining the actions of LINKADD and LINKDEL.

LINKDEL really maintains two linked lists: the list of occupied slots and the list of unoccupied slots that constitute the free space. The program begins by displaying an empty list.

```
List is empty            f→  1 ----  2

Free space starts            2 ----  3

in slot 1                    3 ----  4

                             4 ----  5

                             5 ----  6

                             6 ----  7

                             7 ----  8

                             8 ----  9

Please type                  9 ---- 10

A to add a name,            10 ---- 11

D to delete a name,         11 ---- 12

T to trace a list,          12 ----  0

Q to quit:
```

The annotation "free space starts in slot 1" at the top left of the screen shows that slot 1 is where the next new item will be stored. Slot 1 is linked to slot 2, which is linked to slot 3, and so on to slot 12. It is linked to a fictional slot 0, showing that it is the end of the free space. If you type A to add a name, the program prompts you to type it. Thus LINKDEL begins in much the same way as LINKADD. You enter a name, for example, Al, and the program prompts you to press the space bar to store the new name in slot 1. The program then follows the step-by-step progression of prompts and color changes described for LINKADD. Al is installed in slot 1. This is now both the head and the tail of the list. The free space starts in slot 2. The top of the display changes to:

```
Head of list is          h→  1 Al    0

in slot 1                f→  2 ----  3

Free space starts            3 ----  4

in slot 2                    4 ----  5

...
```

The prompt to type A, D, T, or Q reappears at the bottom left of the screen. Type A and enter another name, say, Hal. Press the space bar to transfer it to slot 2. The second slot of the list changes to:

2 Hal   /

The program appends each new entry to the end of the list, without regard to the other entries. When you respond to the prompts to make this happen, the tie line is drawn:

1 Al   2⌐
─────────────
  └─> 2 Hal   /

Then the link from the new entry is reset to 0 and the tie line is erased. The top of the display now is:

```
Head of list           h→ 1 Al    2

is in slot 1              2 Hal   0

Free space starts      f→ 3 ----  4

in slot 3                 4 ----  5
```

The prompt to type A, D, T, or Q reappears. You can continue to add new names in the manner that has been described. Each new name is transferred to the slot that currently starts the free space. The name in the preceding slot is linked to it, regardless of alphabetic order. The link from the slot containing the new name is reset to 0, and the start of the free space moves to the next slot in the array. After several cycles, the display might be as follows:

```
Head of list           h→ 1 Al    2

is in slot 1              2 Hal   3

Free space starts         3 Bill  4

in slot 6                 4 Carl  5

                          5 Sal   0

                       f→ 6 ----  7

                          . . .

                         12 ----  0
```

Slot 5 now is the tail of the list, and slot 12 is the tail of the free space. Now type D. The program prompts for the name to be deleted. Enter the name Hal. It is displayed for reference on the left of the screen. You are prompted to press the space bar to start searching. This begins a cycle that is rather like the search cycle of the LINKADD program. Prompts appear; the colors of entries under attention change; and tie lines are drawn and erased. Thus, after finding that the name to be deleted does not match the entry at the head of the list, the tie line is drawn to the next entry.

```
    1  Al      2┐
   ┌─          ─│
   └─> 2  Hal     3

       3  Bill    4
```

When the program compares the name to be deleted with the name in slot 2, it finds that they match. Accordingly, it states so in the next prompt and then prompts you to link the predecessor of the selected item to the latter's successor. You respond, and the program:

1. Changes the link in the first slot to 3

2. Draws the tie line from slot 1 to slot 3

3. Changes the link in slot 2 to a slash to show it is in a temporary state of limbo:

```
    1  Al      3┐
                │
    2  Hal      / │
   ┌─           │
   └─> 3  Bill    4
```

The colors of the entries under attention also change to focus attention on these actions. Then the program prompts you to press the space bar to link the old tail of the free space to the slot left by the deleted item. You respond and the program:

1. Erases the tie line from slot 1 to slot 3

2. Changes the entry in slot 2 to ---- and the link to 0 to show it is the new tail of the free space

3. Redisplays slot 12 with the link changed to 2

4. Draws the tie line from the old tail of the free space in slot 12 to the new tail in slot 2

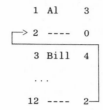

```
        1  Al     3
   ┌─> 2  ----    0
        3  Bill   4
         . . .
       12  ----   2┘
```

After the next key press, the tie line is erased, and the slots 1 and 6 are marked as head of the list and the start of the free space, respectively. If you proceed to add more items, they go into slots 6, 7, ... Only when slot 12 has been filled does a new name go into slot 2.

Making the freed slot the new tail of the free space is not the only way to design the deletion part of a linked list maintenance program. As an alternative, the freed slot can be made the new start of the free space (see Suggested Activities). Most of the time, the effects of the choice are invisible and inconsequential. The visibility here results from the explicit purpose of LINKDEL to show what is happening. Putting the freed slot at the end of the free space keeps the items in their order of arrival until wraparound occurs. A similar situation is described for a circular queue in Section 4.6. Also, there are physical situations that warrant actions equivalent to putting the freed slot at the end of the free space. For example, if the chairs in a waiting room are taken to be the analogy of slots in the linked list, the corresponding tactic tends to distribute the wear and tear of upholstery in a uniform fashion. If parallel units of an electronic system are identified with the slots, the tactic distributes the use of the units in a uniform way that can be significant when the units are subject to random failures.

In general, the following situations can occur when a deletion is requested.

1. The head item of the list is deleted, which leaves one or more items in the list.

2. An internal item is deleted; that is, the deleted item is neither the head nor the tail.

3. The tail of the list is deleted, which leaves one or more items still in the list.

4. The list contains only one item, and that item is deleted.

5. The list does not contain the item that is specified in the deletion request.

Case 1 is illustrated by deleting Al from the following list:

```
Head of list              h→  1  Al    2

is in slot 1                  2  Hal   3

Free space starts             3  Sal   0

in slot 4                 f→  4  ----  5

Delete Al                     5  ----  6

. . .                           . . .

                             12  ----  0
```

The result is:

```
Head of list                  1  ----  0

is in slot 2              h→  2  Hal   3

Free space starts             3  Sal   0

in slot 4                 f→  4  ----  5

                                . . .

                             12  ----  1
```

Case 2 is illustrated by the example, given earlier, in which Hal was deleted from a list that began with Al, Hal, Bill, ... Case 3 is illustrated by deleting Sal from the following list:

```
Head of list              h→  1  Al    2

is in slot 1                  2  Hal   3

Free space starts             3  Sal   0

in slot 4                 f→  4  ----  5

Delete Sal                    5  ----  6

. . .                           . . .

                             12  ----  0
```

The result is:

```
Head of list          h→  1  Al      2

is in slot 1              2  Hal     0

Free space starts         3  ----    0

in slot 4             f→  4  ----    5

. . .                        . . .

                         12  ----    3
```

Case 4 is illustrated by deleting Al from the list shown in the display:

```
Head of list          h→  1  Al      0

is in slot 1          f→  2  ----    3

Free space starts         3  ----    4

in slot 2                 4  ----    5

Delete Al                 5  ----    6

. . .                        . . .

                         12  ----    0
```

The result is:

```
List is empty             1  ----    0

Free space starts     f→  2  ----    3

in slot 2                 3  ----    4

                             . . .

                         12  ----    1
```

Because the slot freed by a deletion is linked to the tail of the free space, the display is not restored to its original condition even though the list is empty. The free space can be connected in any of 12! different arrangements by a suitable history of insertions and deletions that lead to an empty list.

When the list does not contain the name to be deleted, the search is ended by a prompt that tells you this and allows you to proceed. Fur-

```
100 REM * LINKDEL deletes from an unordered linked list *
101 :
        ... statements 110-560 are the same as in LINKADD
570    IF N<NLIM THEN P(N)=N+1 ELSE P(N)=0                          'link
        ... statements 580-620 are the same as in LINKADD
630    FTAIL=NLIM                                          'tail of free space
        ... statements 999-1070 are the same as in LINKADD
1080   IF NF=0 THEN PRINT "List is full."
        ELSE PRINT "Free space starts" S$ "in slot" NF         'free space
1090   VCP=CSRLIN+2
1100   IF P(0)>0 THEN LOCATE 2*P(0)-1,HCP-2: PRINT "h" RA$
1110   IF NF>0 THEN LOCATE 2*NF-1,HCP-2: PRINT "f" RA$
1120   K=1: GOSUB 6500                                     'prompt for action
1130   R$=INPUT$(1)
1140   OP=INT((INSTR("AaDdTtQq",R$)+1)/2)                         'op code
1150   IF OP=0 THEN OP=1                                          'default
1160   LOCATE 1,1,0,7
1170   ON OP GOTO 2000,3000,4000,12000                            'branch
1999 :
2000 REM * insert a new item into the list *
2010   IF NF>0 THEN 2050
2020   K=20: GOSUB 6500                          'no room for another item
2030   Z$=INPUT$(1)
2040   GOTO 1000
2050   K=2: GOSUB 6500                              'prompt for the item
2060   LOCATE CSRLIN,POS(0),0,7                        'to be inserted
2070   INPUT; " ",A$
2080   K=3: GOSUB 6500                    'prompt for key press to store it
2090   PRINT NF;
2100   NS=NF                                             'set the slot #
2110   Z$=INPUT$(1)
2120   A$(NS)=A$                          .              'store the new item
2130   NF=P(NS)                                     'reset start of free space
2140   LOCATE VCP-3,8                               'redisplay its value
2150   PRINT NF
2160   LOCATE 2*NS-1,HCP-2                             'erase its marker
2170   PRINT "  "
2180   IF P(0)=0 THEN LOCATE 1,6: PRINT "was empty": GOTO 2230
2210   LOCATE 2*P(0)-1,HCP-2                            'erase its marker
2220   PRINT "  "
2230   P(NS)=-1
2240   COLOR 7,0
2250   GOSUB 6020
2400   N=TAIL                   'prepare to link old tail to new item
2550   P(NS)=0                                   'new item is new tail
2560   TAIL=NS
2610   K=12: GOSUB 6500                     'prompt to link to new item
2660   Z$=INPUT$(1)
2680   P(N)=NS                                  'reset link to new item
2690   GOSUB 6020                                   'redisplay the item
2700   M=N: GOSUB 7000                            'display the tie line
2710   IF N>0 THEN GOSUB 6010              'de-highlight the linked item
2730   K=13: GOSUB 6500                            'prompt to continue
2740   Z$=INPUT$(1)
2750   GOSUB 7500                              'erase the tie lines
2780   GOTO 1000
2999 :
3000 REM * prompt for an item to delete *
3010   IF P(0)>0 THEN 3050
3020   K=21: GOSUB 6500                               'cannot delete from
```

```
3030    Z$=INPUT$(1)                                                    'empty list
3040    GOTO 1000
3050   K=2: GOSUB 6500                                          'prompt for the item
3060    LOCATE CSRLIN,POS(0),0,7                                   'to be deleted
3070   INPUT; " ",A$
3080    LOCATE VCP,1,0,0
3090     PRINT "Delete ";                                         'display the item
3100     COLOR 13,0                                                 'to be deleted
3110     PRINT A$
3120     COLOR 7,1
3130    IF P(0)>0 THEN LOCATE 2*P(0)-1,HCP-2: PRINT " "              'erase h and
3140   IF NF>0 THEN LOCATE 2*NF-1,HCP-2: PRINT " "                   'f markers
3150   GOSUB 5000                                                'search the list
3160   IF NOT UNPLACED THEN 3500                           'jump if item was found
3170    K=14: GOSUB 6500
3180    Z$=INPUT$(1)                                             'it was not found
3190    GOSUB 6010                                            'dehighlight the tail
3200    GOTO 1000
3499  :
3500 REM * reset the links *
3510    K=15: GOSUB 6500                                     'display message that
3520    Z$=INPUT$(1)                                            'item was found
3530    ND=P(N)                                                'set its slot #
3540    IF N=0 THEN K=18 ELSE IF P(ND)=0 THEN K=16 ELSE K=17         'prompt for
3550     GOSUB 6500                                              'next action
3560    Z$=INPUT$(1)
3570    GOSUB 7500                                              'erase tie line
3580    P(N)=P(ND)                               'link from predecessor to successor
3590    P(ND)=-1                                 'undefine link from deleted item
3600     COLOR 13,0                                             'redisplay it
3610    GOSUB 6050
3620    IF N>0 THEN GOSUB 6010                           'dehighlight predecessor
3630   IF P(N)>0 THEN 3660                     'jump unless deleted item was the tail
3640    TAIL=N                                         'reset tail to predecessor
3650     IF N>0 THEN GOSUB 6040                             'redisplay it
3660    M=N: GOSUB 7000                                         'draw tie line
3670    K=19
3680     GOSUB 6500                                           'prompt to link free
3690     Z$=INPUT$(1)                                        'space to freed slot
3700     GOSUB 7500                                             'erase tie line
3710    P(FTAIL)=ND                        'reset link from previous tail of free space
3720    P(ND)=0                                   'make the freed slot the new tail
3730    A$(ND)="----"                                          'nullify the entry
3740     NN=FTAIL: GOSUB 6200                         'redisplay the previous tail
3750    M=FTAIL: GOSUB 7000                                     'draw tie line
3760    GOSUB 6050                                      'display the nullified entry
3770    K=13: GOSUB 6500                                         'prompt to continue
3780     Z$=INPUT$(1)
3790    GOSUB 7500                                             'erase tie line
3800   FTAIL=P(FTAIL)                                    'reset tail of free space
3810   GOTO 1000
        ... statements 3999-5130 are the same as in LINKADD
5140    IF OP=2 THEN K=7 ELSE 5190                          'prompt for next action
5150     GOSUB 6500
5160     Z$=INPUT$(1)
5170    IF OP=2 THEN UNPLACED = (A$(P(N))<>A$)                          'compare
5180    IF NOT UNPLACED THEN 5260                              'jump if match
5190    IF OP=2 THEN K=9 ELSE K=13                         'prompt for next action
        ... statements 5200-6030 are the same as in LINKADD
6040   NN=TAIL: GOSUB 6200: RETURN                               'slot for tail
```

```
6Ø5Ø   NN=ND:   GOSUB 62ØØ: RETURN                    'slot for deleted item
     ... statements 62ØØ-7999 are the same as in LINKADD
8ØØØ REM * actual messages *
8Ø1Ø   DATA 5,  "Please type:", "A to add a name,", "D to delete a name,",
                "T to trace a list,", "Q to quit:"                     :'1
8Ø2Ø   DATA 3,  "Type a name",   "(limit is 4", "letters):"            :'2
8Ø3Ø   DATA 4,  "Press the space", "bar to store the", "new name in", "slot"  :'3
8Ø4Ø   DATA 3,  "Press the space", "bar to start", "searching"         :'4
8Ø5Ø   DATA 3,  "Press the space", "bar to find the", "next entry"     :'5
8Ø6Ø   DATA 1,  "dummy message for consistent numbering"               :'6
8Ø7Ø   DATA 5,  "Press the space", "bar to compare", "the name to be",
                "deleted with the", "highlighted entry"                :'7
8Ø8Ø   DATA 1,  "dummy message for consistent numbering"               :'8
8Ø9Ø   DATA 4,  "The names are", "different - press", "the space bar to",
                "continue"                                             :'9
81ØØ   DATA 1,  "dummy message for consistent numbering"               :'1Ø
811Ø   DATA 1,  "dummy message for consistent numbering"               :'11
812Ø   DATA 4,  "Press the space", "bar to make the", "new entry into",
                "the new tail"                                         :'12
813Ø   DATA 2,  "Press the space", "bar to continue"                   :'13
814Ø   DATA 5,  "The item to be", "deleted is not", "in the list -",
                "press the space", "bar to continue"                   :'14
815Ø   DATA 3,  "The items match -", "press the space", "bar to continue"  :'15
816Ø   DATA 5,  "Press the space", "bar to make the", "predecessor of",
                "the selected", "item the tail"                        :'16
817Ø   DATA 5,  "Press the space", "bar to link the", "predecessor of",
                "the selected item", "to its successor"                :'17
818Ø   DATA 5,  "Press the space", "bar to make the", "successor of",
                "the selected item", "the new head"                    :'18
819Ø   DATA 6,  "Link the old", "tail of the", "free space to", "the slot",
                "left by the", "deleted item"                          :'19
     ... statements 82ØØ-12Ø7Ø are the same as in LINKADD
```

ther self-explanatory messages are displayed if you try to delete an item from an empty list or add an item when the list is full.

Program Variables

The program LINKDEL uses the variables that occur in LINKADD and also the following:

FTAIL   The slot number of the tail of the free space

VCP   The vertical position of the cursor, used to position the name that is to be deleted

ND   The slot number of the item to be deleted, when it has been found

Program Notes

The general structure of LINKDEL is very similar to that of LINK-ADD. The initialization section of LINKDEL repeats the corresponding actions of LINKADD. It also sets FTAIL to 0 and links the successive slots to form the initial free space.

Following the main prompt, OP is set to the values 1, 2, 3, 4 to add, delete, trace, and quit respectively. Then the program branches accordingly. The action to add an item is much simpler in LINKDEL than in LINKADD. The new item is just put into the first available slot and made the new tail. Instead of using the search subroutine to find the value of N to link to the new name, a few simple assignments are used to link the old tail to the new name and to identify the new name as the new tail.

```
2400   N=TAIL
2550   P(NS)=0
2560   TAIL=NS
2680   P(N)=NS
```

The search subroutine is used for the delete action and the trace action. The only difference from LINKADD is in the condition that sets the switch UNPLACED in the WHILE . . . WEND loop that works through the list. This statement is now:

```
5170 IF OP=2 THEN UNPLACED = (A$(P(N))<>A$)
```

If the requested item matches the item at the head of the list, that is, the item in slot P(0), then UNPLACED is set to 0 on the first passage through the loop. The statement to update N is bypassed, and the next execution of the WHILE statement sends control past the end of the loop. If the requested item does not match the item at the head of the list, N is reset to point to the next item. The program then cycles until either (1) P(N) points to the matching item when the comparison is made or (2) the end of the list is reached without finding a match. After leaving the loop, in case 2, UNPLACED still is −1 and P(N) is zero.

Following the WHILE . . . WEND loop, the program branches on UNPLACED to deal with the cases that the item to be deleted has or has not been found. A diagnostic prompt is displayed when the item has not been found. When it has been found, the links are updated by the statements:

```
3530   ND=P(N)
. . .
3580   P(N)=P(ND)
. . .
3630   IF P(N)>0 THEN 3660
3640   TAIL=N
. . .
3660   . . .
. . .
3710   P(FTAIL)=ND
3720   P(ND)=0
. . .
3800   FTAIL=P(FTAIL)
```

The further statements in the section to reset the links provide the visual effects. The three cases that require separate prompts are distinguished as follows:

N = 0	The head is to be deleted and the successor of the matching item is made the new head. Strictly speaking, this applies when the list contains at least one more entry, that is, P(ND)>0. When the head is to be deleted and it is the only item in the list, that is, N = 0 and P(ND) = 0, then the list is cleared.
N>0 and P(ND)>0	An internal item is to be deleted—the predecessor of the matching item is linked to the successor.
N>0 and P(ND) = 0	The tail is to be deleted—the predecessor of the selected item is made the new tail.

Suggested Activities

1. Adapt LINKDEL to display an expanded main menu and, when requested, to delete:
   (a). The successor of a given name
   (b). The predecessor of a given name
   (c). The name that occurs in the nth position alphabetically
   (d). All occurrences of a given name
   (e). All the entries that precede a given name
   (f). All the entries that follow a given name
   (g). All the entries between two names that you type in response to further prompts

2. Expand LINKDEL to demonstrate the deletion of items from:
   (a). A doubly linked list
   (b). A circular linked list
   (c). A circular doubly linked list

3. Expand LINKDEL to maintain n linked lists in the same array. When a name is added or deleted, specify which list should be altered.

4. Alter LINKDEL to put the freed slot at the head of the free space.

3.3   Letting an Ordered List Wax and Wane

The program LINKLIST combines the actions of LINKADD and LINKDEL. The initial prompt is the same as in LINKDEL. Thereafter, the addition of a name proceeds in the manner described for LINK-

```
100 REM * LINKLIST - inserts and deletes in a linked list *
101 :
        ... statements 110-2180 are the same as in LINKDEL
2190    LOCATE 1,14                                         'alter the
2200    PRINT "was"                               'comment about the head
        ... statements 2210-2250 are the same as in LINKDEL
2400    GOSUB 5000                                  'search the linked list
2499 :
2500 REM * reset the links *
2510    IF P(N)=0 THEN 2550                  'jump if new item is new tail
2520    K=10: GOSUB 6500
2530    Z$=INPUT$(1)                     'prompt to link predecessor of
2540    GOSUB 7500                         'item in slot N to new item
2550    P(NS)=P(N)                         'set link from new item
2560    IF P(N)=0 THEN TAIL=NS: GOTO 2610      'new item is new tail
2570    M=NS: GOSUB 7000                  'draw tie line from new item
2580    COLOR 7,0                          'de-highlight the item
2590    GOSUB 6020
2600    IF N=0 THEN 2680                     'jump if new item is new head
2610    IF P(N)>0 THEN K=11 ELSE K=12: GOTO 2650  'prompt to link to new item
2620    P(N)=-1                            'undefine link in slot N
2630    COLOR 3,0                          'redisplay contents of slot N
2640    GOSUB 6010
2650    GOSUB 6500                                   'display the prompt
2660    Z$=INPUT$(1)
2670    GOSUB 7500                                   'erase the tie line
2680    P(N)=NS                            'reset link to new item
2690    GOSUB 6020                         'redisplay the item
2700    M=N: GOSUB 7000                    'display the tie line
2710    IF N>0 THEN GOSUB 6010           'de-highlight the linked item
2720    IF P(NS)>0 THEN NN=P(NS): GOSUB 6200   'de-highlight the successor
2730    K=13: GOSUB 6500                    'prompt to continue
2740    Z$=INPUT$(1)
2750    GOSUB 7500                         'erase the tie lines
2760    M=NS
2770    GOSUB 7500
2780    GOTO 1000
2999 :
        ... statements 3000-3150 are the same as in LINKDEL
3160    IF NOT UNPLACED AND A$=A$(P(N)) THEN 3500     'jump if item was found
3170    K=14: GOSUB 6500
3180    Z$=INPUT$(1)                               'it was not found
3190    IF N=TAIL THEN GOSUB 6010 ELSE GOSUB 6030  'dehighlight item it passed
        ... statements 3200-5130 are the same as in LINKDEL
5140    IF OP<3 THEN K=OP+5 ELSE 5190               'prompt for next action
5150      GOSUB 6500
5160      Z$=INPUT$(1)
5170    IF OP<3 THEN UNPLACED = (A$(P(N))<A$)                 'compare
5180    IF NOT UNPLACED THEN 5260                   'jump if match
5190      IF OP<3 THEN K=OP+7 ELSE K=13             'prompt for next action
        ... statements 5200-8050 are the same as in LINKDEL
8060 DATA 5, "Press the space", "bar to compare", "the new name with",
        "the highlighted", "entry"                             :'6
8070 DATA 5, "Press the space", "bar to compare", "the name to be",
        "deleted with the", "highlighted entry"                :'7
8080 DATA 5, "The new name is", "alphabetically", "higher - press",
        "the space bar", "to continue"                         :'8
8090 DATA 5, "The name to be", "deleted is higher", "alphabetically -",
        "press the space", "bar to continue"                   :'9
8100 DATA 4, "Press the space", "bar to link the", "new entry to the",
        "highlighted entry"                                    :'10
8110 DATA 6, "Press the space", "bar to link the", "predecessor of",
        "the highlighted", "entry to the", "new entry"         :'11
        ... statements 8120-12070 are the same as in LINKDEL
```

ADD. The deletion of a name begins in the manner described for LINKDEL. If the name that was specified is not in the list, however, the search ends as soon as an alphabetically higher name is reached.

LINKLIST uses the same variables as LINKDEL. The sections that are common to LINKADD and LINKDEL are used again, without change. The section to add items is copied from LINKADD. Within the search subroutine, the statement that performs the critical test has been changed to

```
5170 IF OP<3 THEN UNPLACED = (A$(P(N))<A$)
```

In the section to delete a name, which is copied from LINKDEL, the statement that follows the execution of the search subroutine has been changed to:

```
3160  IF NOT UNPLACED AND A$=A$(P(N)) THEN 3500
```

These two changes permit an unsuccessful search to end as soon as the required alphabetic position has been passed.

Suggested Activities

1–7. Adapt the LINKLIST program along the lines of Suggested Activities 1 to 7 of Section 3.1, singly or in combination.

8–10. Adapt LINKLIST along the lines of Suggested Activities 1 to 3 of Section 3.2, singly or in combination with each other and with Activities 1 to 7.

11. To reinforce the logic that is used to end an unsuccessful search, alter LINKLIST to use a variable PLACED instead of UN-PLACED. [*Hint:* Initialize PLACED to zero and reset it to A$>= A$(P(N)) in line 5170. Change line 3160 to IF PLACED AND A$ = A$(P(N)) . . . .]

3.4 Some Benefits of Linked Lists

Mention must be made, as a preliminary, of some details of terminology. We have used the word "pointer," informally, for the subscript of the array item to which attention is being directed. The word is expressive and convenient for this purpose, and it was used without any more precise force for several years. A linked list can be described in a way that is a little more general by considering it to be a collection of *nodes* each of which consists of an *information part* and a *next address part*.

This escapes the specificity of using an explicitly declared array. It alludes, however, to the idea of storage address, which users of high-level languages in general avoid.

PASCAL, PL/1, and some other languages do include provisions for "pointer variables" that are assigned, in effect, the addresses of other items of data. The data then can be stored and retrieved by reference to the pointer variables. These are a convenience in the implementation of linked list programs. Some authors use the word "pointer" with the restricted meaning of "pointer variable" and use "cursor" where we have used "pointer." We have stayed with the informal use of "pointer" because we have to use the word "cursor" for the position of text on the screen.

The programs and suggested activities in this chapter demonstrate five major uses of linked lists that can be made independently or in combination, with a number of possible benefits.

1. The representation of order can be maintained in a volatile situation, without the continual physical rearrangement of data.

2. Different conceptual arrangements of the same set of records, defined by reference to data in different fields, can be maintained concurrently.

3. Several collections of data can be built in the same storage area, subject to a preset limit on just their combined size.

4. Space that is vacated by deletions can be reused.

5. Several "dramatic" operations, such as deleting a substantial portion of a list or switching a sequence of elements from one list to another, can be implemented concisely.

The first of these avoids the physical rearrangement of data at the cost of (1) the space occupied by the links and (2) the time taken in the entry-by-entry search to find where a new item fits in sequence. The search time can create serious problems when the list is very long, but this effect can be alleviated by splitting the list into several consecutive pieces of comparable size. Before adding a new entry, a preliminary step determines into which sublist it should be inserted. Thus, if roughly equal numbers of names begin with each letter of the alphabet, 26 sublists can be maintained, and each new name is inserted in the sublist that is determined by the first letter of the name. This reduces the average search time by a factor of 26. Also, if a doubly linked list is used, searching can be made to proceed backward from the end of the list when, by reference to an appropriate criterion, the new item seems closer to the end than to the beginning.

The second use of linked lists, mentioned above, provides an alternative to the use of multiple indexes that was described in Section 2.4. In a volatile situation, maintaining the multiple indexes serially involves continual rearrangement of the entries. The trade-off between that approach and the use of linked lists is similar to the one discussed in the preceding paragraph. Moreover, each of the multiple linked lists can be split into separate sublists. The membership of these is quite independent.

Thus, in the bibliography example that follows, the first and second links pertain to book titles and to authors' names, respectively. Each linked list is split into two parts for the initial letters A to M and N to Z. The heads of the two halves of the author sublists and the two halves of the title sublists are in slots 1 to 4, respectively.

Record number	Author name	Author link	Title	Title link
1	Adams	4	Trash	5
2	Nash	5	Crash	6
3	White	0	Bunk	2
4	Brown	6	Shrunk	1
5	Trent	3	Tripe	0
6	James	0	Gripe	0

Thus, the first author sublist consists of records 1, 4, and 6, and the second of records 2, 5, and 3. The first title sublist consists of records 3, 2, and 6, and the second of records 4, 1, and 5. The third use of linked lists mentioned above has an alternative when only two lists are built in a single array. One sequential list can be stored upward from the bottom of the array. The other is stored downward from the top.

The fourth use of linked lists is that the links facilitate the reuse of freed space. There are several alternatives to using linked lists for this purpose when the links are not also used to show the alphabetical or numerical order of entries. A special value can be put into each slot initially to show that the slot is empty. Then, when an item is deleted, the content of the slot is reset to this "empty indicator." A free slot can be found by searching sequentially for a slot that contains the empty indicator. In contrast, linking the unoccupied items avoids the need to test the content of slots that happen to be occupied when searching for space to put a new item. Linking the occupied slots avoids the need to test the content of slots that happen to be unoccupied when searching for an item that must be deleted. Another way to keep track of slots that are occupied and unoccupied without using linked lists is to use a bit string. The nth bit is 0 if the nth slot is empty. The bit is 1 if the slot is occupied. This "monitor string" is initialized to zeros. If its name is M\$, then the subscript of the first available slot is given in BASIC by INSTR(M\$,"0"). The bit for the nth slot can be changed by assigning 1 or 0 to MID\$(M\$,n,1). The subscript of the first occupied slot after slot k is INSTR(k,M\$,"1").

For the fifth kind of use mentioned above—"dramatic" changes to a list—linked lists have strong advantages over other representations. The entries with a particular sequence of subscript values can be deleted from a sequential array by a loop that moves subsequent entries back by an appropriate amount or changes the content of the slots into the empty slot indicator. A single instruction can reset a succession of bits to 0 in a monitor string that shows which members of a set of possible elements are present. Transferring a succession of entries from one list to another, however, requires the physical movement of data when the sequential array representation is used. It can be represented by resetting bits if a separate monitor string is used for each list. When the order of the entries is significant, however, the linked list approach is usually preferable or necessary.

Changes to a list are particularly rapid and frequent when the list represents the successive interim results of an ongoing algorithm. Thus, the manipulation of mathematical expressions, such as differentiating a multinomial, creates lists of symbols that must be kept in order and which wax and wane very rapidly.

Once the decision to use a linked list has been made, a further trade-off occurs when order is not of concern to the user. Maintaining links that show the alphabetic or numeric relations of the entries lets you end the search for an item that is to be deleted as soon as an entry with a higher value is reached. Otherwise, the search must go to the end of the list whenever the value specified for deletion is absent. Suppose, however, that some details of the application ensure that requests for deletion are always valid and are distributed randomly through the list. Then half the entries have to be tested on average, whether they are approached sequentially or at random. Maintaining links to show the alphabetical or numerical order of the items then slows the insertion process considerably without expediting the deletion.

This kind of analysis can be extended to several other situations that pertain to ordinary linked lists (of the kind discussed in most of this chapter), to doubly linked lists (in which each item is accompanied by a link to its predecessor as well as to its successor), and to circular linked lists (in which the tail is linked to the head). Linked list operations thus provide considerable scope for varied design and analysis.

Linked lists can also be used, in conjunction with serial addressing, for storage control in a way that the following word processing application illustrates. Suppose that the successive lines of a draft text are stored as successive elements of an array. In particular, suppose that the draft is 100 lines long and that lines 20 to 30 are to be swapped with lines 50 to 60. The revised text consists of lines 1 to 19, 50 to 60, 31 to 49, 20 to 30, and 61 to 100, in that order. The revised text can be represented by the unchanged content of the original array and a short list of ranges of subscript values that point to it. New text to be inserted can be put into the primary array at the end of the text that was stored initially if note is kept of where the new lines fit. The effects of deletions and insertions, as well as transpositions, can be represented, accordingly, by maintaining a list of ranges of subscript values. This list can be stored sequentially in a second array. Alternatively, it may be advantageous to represent it by a linked list when frequent changes are made. Numerous variations on this basic theme are made when a large body of data consists of blocks that can be stored serially. In some cases, ranges of actual storage addresses in high-speed memory or on disk are used. Sometimes a directory simply points to the first (or only) block in a body of data and each block ends either with a code that shows it is not continued or with a link to the next block.

3.5 Shielding the User from the Gory Details

Most of the emphasis of the earlier sections has been on programming techniques to deal with lists and on displays that make the techniques very visible. Suppose, however, that you are given a program that lets you create and operate on lists as follows. If l stands for a name that you want to give to a list that you are going to create, then typing the command N,l creates a null list with that name. A null list contains no elements; its length is 0. Suppose, too, that the program lets you insert the item v in the alphabetically correct position of the list l by typing the command I,v,l. To create a list l that contains the items v_1, v_2, . . . , v_m in alphabetic order, you type the commands N,l; I,v_1,l; I,v_2,l; . . . ; I,v_m,l. Suppose that the program deletes the item v from the list l when you type the command D,v,l. Suppose also that the program prints the list l when you type the command P,l.

If you have an application that requires the insertion and deletion of items from lists that are printed in alphabetic order intermittently, this program meets your need. You need not know whether the list is maintained as a linked list or as a sequential array or in yet some other way. It is enough to know that the program keeps track of the alphabetic order of the elements or permits it to be found when necessary.

When we deal with a body of information in this way, without concern for the internal details of how the computer stores and processes it, then we are said to treat the information as an abstract data type (ADT). An ADT is characterized by the external effects of the operations that can be performed on it. When dealing with lists, we take for granted a sequencing criterion that can be used to put a set of elements in order. A list is completely defined if we can find (1) its length, (2) the element at position k in its ordered arrangement, where k takes values from 1 to its length. Denote the length of list l by $L(l)$ and the kth element by $E(k,l)$. Starting with these two primitives, many more properties can be defined. Thus, the position of the first element with value v is the lowest value k for which $E(k,l) = v$ is true.

We can regard finding the length and finding the kth element as two of the basic operations that can be performed on lists. The results of applying a simple set of operations on an alphabetically ordered list can be expressed in function form as follows:

$N(l)$ Nullifies the list called l. Thus, if a list with this name does not exist, the program creates a null list, that is, a list containing no elements, called l. If a list called l does exist, all its entries are deleted and its length is set to 0.

$I(v,l)$ Inserts the item v into list l in the correct alphabetic position. We assume that this command provides the only method of putting entries into the list and thereby ensuring the creation of the list in the appropriate order.

$D(v,l)$ Deletes the first (or only) occurrence of v from l.

$L(l)$ Obtains the length of l.

$E(k,l)$ Obtains the kth element of l in alphabetic order.

$P(v,l)$ Obtains the position of the first (or only) occurrence of v in l.

$W(l)$ Prints the list in alphabetic order.

These can be expressed in terms of a shorter set of primitives, and they can also be used to build further operations. Thus the successor of v is $E(P(v,l) + 1,l)$. It is convenient to define $E(k,l)$ to be null when $k<1$ and when $k>L(l)$ and to adopt some further related conventions. For practical purposes, a list-processing program should provide the user with simple commands for a much richer set of operations, even though they can be constructed from the primitives.

When dealing with bodies of information like lists, it is important to define convenient sets of operations that are independent of the internal details of the supporting programs. The reason is that many applications can be defined in terms of the operations that serve, in effect, as commands in a specialized high-level language. In the example of the past few paragraphs, the commands can be introduced as immediate mode actions of an interactive package. The concept and the implementation can be extended to allow stored procedures that contain the commands. Applications that are defined in this way can be transported between list-processing systems with the same external specifications and without regard to the internal details. This is particularly important when changes in technology warrant major changes in internal implementation techniques, such as a switch from linked list to sequential array representation or vice versa.

List processing in general, and linked lists in particular, lead on to many other topics. These include, for example, the recursive definition of lists, that is, lists which contain other lists as elements. The literature of list-processing systems is extensive, and sources for further reading are included in the Bibliography.

Suggested Activities

1. Write a program that supports the set of commands N(l), . . . , W(l) described in this section. Use linked list techniques.

2. Write the corresponding program that uses sequential array storage.

3. (a). Extend the programs of Activities 1 and 2 to store and execute sequences of these commands.
   (b). Extend Activity (a) to include conditional statements, branching, and looping.

4. Extend the linked list programs to allow the entries in the lists to be records consisting of several fields. Let the user define (a) the number of fields, (b) which fields are numeric, (c) which fields are alphabetic, and (d) the fields for which links are to be provided to show the relative position of the records in logical order. Expand the I, E, and P commands to let the user specify the sequencing field.

5. Consider the application of the program described in Activity 4 to problems of information retrieval. List further commands that would be useful.

Working with Stacks and Queues

This chapter, in common with Chapters 1 to 3, deals with ordered collections of data. The order in which data is encountered in a dynamic situation is the main concern here, however, rather than the numeric or alphabetic order. In particular, we deal with *stacks* and *queues*. Stacks embody a very familiar pattern of human behavior that consists of continually redirecting attention to the most recent interruption and going back to the matter of prior attention when the latest interruption has been dealt with. In other words, a stack supports a last in, first out (LIFO) mode of operation. It can be compared with the practice whereby trays are taken from the top of a pile and put back on top of the pile in a cafeteria. A queue, by contrast, supports a first in, first out (FIFO) mode of operation. This is just the guiding principle of waiting lines in cafeterias, banks, and innumerable other daily life situations. In fact, the word "queue" is used in England for a waiting line (or, from another viewpoint, the term waiting line is used for a queue of people in the United States).

4.1 Reversing a List

The program REVERSE demonstrates the use of a stack to reverse a list of items in a very simple fashion. It begins by prompting you to type a list of up to 15 items. Then it draws an empty stack and prompts you to press the space bar. If you had typed the items dog, . . . , spoon shown in the next display, the appearance of the main part of the screen at this point would be:

```
dog, cat, chair, lamp, table, cup, spoon
Press space bar to find the 1st item.
```

You press the space bar. The first item is highlighted, and you are prompted:

```
Press space bar to push current item onto stack.
```

You respond, and "dog" appears in the stack. The word "push" is used for the action of putting an item onto the top of a stack.

```
|            |
| dog        |
```

The program prompts you to proceed to the next item. You press the space bar. The highlighting moves to the second item in the list. The prompt to push the current item onto the stack reappears. You respond, and cat appears in the stack, above dog.

```
| cat        |
| dog        |
```

The program continues to loop in this way. In each cycle it (1) prompts for a key press to find the next item, (2) moves the highlighting to the next item in the list when you respond, (3) prompts for another key press, and (4) pushes the highlighted item onto the stack when you press the space bar again.

In due course, the program reaches the end of the list. The final item is pushed onto the top of the stack, and the prompt states that the end has been reached. The screen has the following appearance.

```
dog, cat, chair, lamp, table, cup, spoon
End reached - press space bar.
                                  |
                        spoon
                        cup
                        table
                        lamp
                        chair
                        cat
                        dog
```

You press the space bar again and the prompt changes to:

```
Press space bar to pop an item.
```

You respond. The word "spoon" moves from the top of the stack to the line below the prompt, that is used to display the output of the reversing process. The prompt stays on the screen.

```
dog, cat, chair, lamp, table, cup, spoon
Press space bar to pop an item.
spoon,
```

```
cup
table
lamp
chair
cat
dog
```

The word "pop" is used for the action of removing the item from the top of the stack. You continue to press the space bar. Each time, the item at the top of the stack disappears from the stack and appears at the end of the output list. Thus, after another four key presses, the screen has the following appearance.

```
dog, cat, chair, lamp, table, cup, spoon
Press space bar to pop an item.
spoon, cup, table, lamp, chair,
```

```
cat
dog
```

After another two key presses, for the example at hand, the stack is empty. The output stream consists of the items in the input list, but in the reverse sequence. Then you can run another demonstration or quit.

```
dog, cat, chair, lamp, table, cup, spoon
Press space bar to continue, Esc to quit.
spoon, cup, table, lamp, chair, cat, dog
```

If the list contains more than 15 items, the diagnostic message:

```
Stack overflow. Type T to truncate, Q to quit.
```

appears when the program tries to push the sixteenth item onto the stack. This lets you reverse the 15 items that were stacked or quit immediately.

The pushing and popping to and from the top of the stack are the essential features of stack use. Several direct applications of the reversal of a list are the bases of suggested activities at the end of the section. The need to reverse lists and strings occurs in innumerable circumstances.

Program Variables

SC	The capacity of the stack
STACK$(k)	The kth item in the array used to store the data in the stack
VCSB	The screen position of the base of the stack
HCS	The screen position of the left edge of the stack
WDS	The width of the stack
BAR$	The graphics character used to draw the sides of the stack
HZL$	The graphics character used to draw the base of the stack
VCP	The vertical position of the text cursor used for prompts and for input
U$	The input list, stored as a string
LU	The length of the input string
MS$	The string that serves as the current prompt or diagnostic
V	The index of the loop that draws the sides of the stack
TOP	The index of the topmost item in STACK$
Q	The position of the end of the next item in the input during the cycle that stacks the list
QP	The previous value of Q
VCR	The vertical cursor position of the next item to be displayed in the reversed list
HCR	The horizontal cursor position of the item
Z$	The one-byte response to the prompts for a key press
A$	The substring of the input that consists of the item under current attention before any spaces have been removed from the start or end
B$	The corresponding string after the spaces have been removed
C$	The item just popped from the stack
VCI	The vertical cursor position of the start of the input item under current attention
HCI	The horizontal cursor position of the item

Program Notes

The stacking section of the program is controlled by a WHILE . . . WEND loop that ends when the end of the next item coincides with the end of the input. Within this loop, the program jumps to the subroutines at lines 3500 and 5000 to delimit an item and to push it onto the stack, respectively.

```
1030 WHILE Q<LU
. . .
1070    GOSUB 3500
. . .
1120    GOSUB 5000
. . .
1140 WEND
```

The other statements in the loop highlight and de-highlight the item as it moves into and out of attention.

The pushing process is very simple. A diagnostic message is set up that lets you truncate or quit if the stack is full.

```
5010   IF TOP<SC THEN 5040
5020   MS$="Stack overflow. Type T to truncate, Q to quit"
5030   GOSUB 3000
```

In valid circumstances, the subscript TOP is increased by 1 to point to the next available slot in the array STACK$. The current item is assigned to the TOPth slot in the array.

```
5040 TOP=TOP+1
5050   STACK$(TOP)=B$
```

The item is displayed on the screen for explanatory purposes.

The unstacking process is controlled by another WHILE . . . WEND loop that continues until the popping subroutine sets the switch EMPTY to −1 to show that the stack is empty. Within this loop the subroutine at line 6000 pops the item that is at the top of the stack. This item, which the subroutine assigns to C$, is displayed. Also, a comma is displayed after the item, unless popping the item empties the stack.

```
2100 WHILE NOT EMPTY
. . .
2130    GOSUB 6000
. . .
2170      PRINT C$+...
. . .
2200 WEND
```

The popping subroutine resets EMPTY if the stack is empty, that is, if TOP is zero. Otherwise, C$ is set to the TOPth element of the array STACK$ and TOP is decreased by 1.

```
6010   IF TOP=0 THEN EMPTY=-1:  RETURN
6020     C$=STACK$ (TOP)
. . .
6050     TOP=TOP-1
```

The element also is erased from the screen display before TOP is decreased.

Suggested Activities

1. Write adaptations of REVERSE that demonstrate the use of a stack in programs that:
   (a). Read a list of words separated by commas, print the words in reverse order when a semicolon is encountered, and repeat the process until a period occurs,
   (b). Generalize the process described in (a) by first prompting for three characters that then have the effects that the comma, semicolon, and period have in (a), and
   (c). Read a sequence of letters, ended by a period, and print the letters in their input sequence followed by the letters in the reverse sequence—for example, if the input is abcd, the output is abcddcba.

2. Write a program that reads a BASIC program and checks that (a) each FOR statement is balanced by a NEXT statement containing the same loop index and (b) each WHILE statement is balanced by a WEND statement.

3. Write a program that has the same overall effect as the program in Activity 1(a) but which shows the items in the input list flowing along a channel that reaches the top of the stack, dropping into the stack, and then, when the time is ripe, popping up and flowing out through another channel.

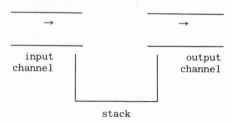

stack

```
100 REM * REVERSE uses a stack to reverse a list *
101 :
110  ON ERROR GOTO 7000                                          'set error trap
120  KEY OFF: SCREEN 0: WIDTH 80
130  COLOR 3,0: CLS                                             'screen set up
140  SC=15                                                   'stack capacity
150  DIM STACK$(SC)                                          'allocate space
160  VCSB=25                                      'position of base on screen
170  HCS=61                                                      'left edge
180  WDS=15                                                          'width
190  BAR$=CHR$(179)                                  'vertical bar character
200  HZL$=CHR$(196)                                          'horizontal bar
210  VCP=6                                              'cursor for messages
299 :
300 REM * type the list *
310  LOCATE 1,1                                                      'prompt
320  PRINT "Please type a list of up to 15 items separated by commas."
330  PRINT "Do not press the Enter key until the end, or use more than 3 lines."
340  PRINT
350  LINE INPUT U$                                               'store list
360  VCP=CSRLIN+1                                        'cursor for messages
370  LU=LEN(U$)                             'length of list stored as a string
380   IF LU<=240 THEN 410                                    'check in range
390    MS$="Too many characters"                                   'too long
400    ERROR 250                                          'trigger error trap
410  MS$="Press space bar to draw the stack"
420  GOSUB 3000                                     'prompt to draw the stack
499 :
500 REM * draw the stack *
510  FOR V=VCSB-SC-1 TO VCSB-1                               'draw the sides
520  LOCATE V,HCS
530  PRINT BAR$ SPACE$(WDS) BAR$;
540  NEXT V
550  LOCATE VCSB,HCS                                          'draw the base
560  PRINT CHR$(192) STRING$(WDS,196) CHR$(217);
570  TOP=0                                             'nullify its contents
999 :
1000 REM * stack the list *
1010  MS$="Press space bar to find the 1st item"         'prompt for 1st item
1020  Q=0                                                      'text cursor
1030  WHILE Q<LU                                        'loop through string
1040   GOSUB 3000                                       'display the prompt
1050   IF Q>0 THEN GOSUB 4000                     'de-highlight unless 1st item
1060   QP=Q                                    'preserve cursor at start of item
1070   GOSUB 3500                            'delimit current item in string
1080    COLOR 10
1090    GOSUB 4000                                            'highlight it
1100   MS$="Press space bar to push current item onto stack"
1110    GOSUB 3000                                          'prompt to push
1120    GOSUB 5000                                  'push item onto stack
1130    MS$="Press space bar to find the next item"     'prompt for next item
1140  WEND
1150  IF TOP>0 THEN GOSUB 4000                         'de-highlight final entry
1999 :
2000 REM * unstack the list *
2010  MS$="End reached - press space bar"           'prompt to start popping
2020   GOSUB 3000
2030   EMPTY=0                                   'set termination switch off
2040   VCR=VCP+2                                  'screen position of current
2050   HCR=1                                           'item of reversed list
```

```
2100   WHILE NOT EMPTY                                      'loop until stack is empty
2110     MS$="Press space bar to pop an item"
2120       GOSUB 3000                                              'prompt to pop
2130       GOSUB 6000                                              'pop an item
2140     IF EMPTY THEN 2200
2150       LOCATE VCR,HCR
2160         IF HCR+LEN(C$)>HCS-2 THEN PRINT               'avoid displayed stack
2170         PRINT C$+MID$(",",(2+(TOP>0)));                'display popped item
2180           VCR=CSRLIN                                   'reset screen position
2190           HCR=POS(0)                                   'for next popped item
2200     WEND
2300     MS$="Press space bar to continue, Esc key to quit"          'prompt
2310       GOSUB 3000
2320       IF Z$=CHR$(27) THEN 8000 ELSE CLS: GOTO 300              'branch
2999 :
3000 REM * display a prompt *
3010   LOCATE VCP,1
3020     PRINT MS$ "." TAB(HCS-1)                          'display a prompt
3030     Z$=INPUT$(1)                                      'wait for key press
3040   RETURN
3499 :
3500 REM * delimit an item *
3510   Q=INSTR(QP+1,U$,",")                                       'find the comma
3520   IF Q=0 THEN Q=LU+1                                         'final item
3530   A$=MID$(U$,QP+1,Q-QP-1)                                    'extract the item
3540   B$=A$
3550   WHILE LEFT$(B$,1)=" "                               'remove spaces at left
3560     B$=MID$(B$,2)
3570   WEND
3580   WHILE RIGHT$(B$,1)=" "                              'remove spaces at right
3590     B$=LEFT$(B$,LEN(B$)-1)
3600   WEND
3610   RETURN
3999 :
4000 REM * highlight or dehighlight an item *
4010   VCI=4+(QP+1)\80                                      'screen cursor position of
4020   HCI=(QP MOD 80)+1                                    'item in input string
4030   LOCATE VCI,HCI
4040     PRINT LEFT$(A$,81-HCI);                            'redisplay to right edge
4050   IF LEN(A$)<=81-HCI THEN 4090
4060     PRINT LEFT$(MID$(A$,82-HCI),80);                  'portion (if any) on next line
4070   IF LEN(A$)<161-HCI THEN 4090
4080     PRINT MID$(MID$(A$,82-HCI),81);                         'and on next
4090   COLOR 3                                             'default color
4100   RETURN
4999 :
5000 REM * push an item onto the stack *
5010   IF TOP<SC THEN 5040                                      'check for overflow
5020   MS$="Stack overflow. Type T to truncate, Q to quit"   'trip error trap
5030     GOSUB 3000
5035     IF Z$="Q" OR Z$="q" THEN 8000 ELSE 2030
5040   TOP=TOP+1                                           'increase stack pointer
5050   STACK$(TOP)=B$                                      'push item onto stack
5060   LOCATE VCSB-TOP,HCS+1
5070     PRINT LEFT$(B$,WDS);                              'display it in place
5080   RETURN
5999 :
6000 REM * pop top item from stack *
6010   IF TOP=0 THEN EMPTY=-1: RETURN                      'set switch if empty
6020     C$=STACK$(TOP)                                    'assign top item
```

```
6030    LOCATE VCSB-TOP,HCS+1
6040      PRINT SPC(WDS)                              'erase it from display
6050      TOP=TOP-1                                   'decrease stack pointer
6060    RETURN
6999 :
7000 REM * error trap *
7010   IF ERR<>250 THEN MS$="Error"+STR$(ERR)+" in line"+STR$(ERL)'error message
7020   MS$=MS$+" - press space bar"
7030   GOSUB 3000                                     'display message
7040   RESUME 8000
7999 :
8000 REM * quit *
8010   LOCATE 23,1
8020   ON ERROR GOTO 0                                'reset error trap
8030   END                                            'quit
```

4. Adapt Activity 3 to deal with an input stream of numbers in the following way. At any time, the number that has arrived at the stack drops in if the stack is empty or if the top item is larger. The number at the top of the stack pops into the output if it is smaller than the next number in the input. Show how this mechanism would deal with a list of numbers that decrease and then increase. Explore possible ways of extending this process to sort a list of numbers, for example, by recycling the list through the mechanism repeatedly.

5. Write a program that uses a stack to convert a statement of the form:

```
take the A of the B of the C of ... of the Z of the x
```

into

```
take the x, form the Z of that, ... form the C of that,
form the B of that, form the A of that
```

Extend this to handle all the prepositions in an array in the program in the same way as "of." Try to find natural-language expressions that a computer would execute by using this tactic.

4.2 Expanding a List

The program INCLIST demonstrates a second major use of stacks: handling digressions. The particular application deals with lists that refer to other lists. Thus, the default demonstration displays six lists called A, B, C, D, E, and F. Each item in these lists is either a word (such as "dog") or a reference to another list. The # is put in front of the name of the referenced list. Thus, #A stands for the elements of A, #B stands for the elements of B, and so on.

```
A:  dog, chair, pipe, rug, cup, book
B:  hat, coat, tie, scarf, shoes
C:  keys, #A, cat, bird, #B
D:  tree, car, #A, bus, #C
E:  train, plane, #A, #C, table
F:  #A, #C, boat, lamp, #E, #A
```

The definitions of A, B, and C make list C, when fully resolved, consist of the elements:

```
keys, dog, chair, pipe, rug, cup, book, cat, bird, hat, coat, tie,
scarf, shoes
```

This is constructed by substituting the elements of A in place of #A and the elements of B in place of #B, in the definition of C.

INCLIST begins by prompting you to type P for preset data, I to input data, or Q to quit. If you type P, the program displays the definition of the six lists A to F, draws an empty stack that is divided into two columns, and prompts for the name of the main list. If you enter the name A, the entire list A is intensified on the screen and you are prompted to press the space bar for the first item. You respond. The word "dog" is colored for further emphasis, and it is copied onto the bottom line of the screen. Then you are prompted to press the space bar for the next item. You respond. The word "dog" is deemphasized. The next item, chair, is colored and copied on the bottom line, following dog. The program continues in this way through the elements of A. In each cycle, it prompts for the next item. When you press the space bar, the coloring moves on to the next item, which also is copied into the output line. The processing of list A continues uneventfully and puts the items dog, chair, pipe, rug, cup, book into the final output line.

Then the program prompts you to type K to keep the present lists or M for the main menu. Respond by typing K. The output line is erased, and the prompt for the main list reappears. This time enter the name C. The list C is highlighted. The program prompts for a key press. The word "keys" is colored and copied onto the output line. After you respond to the next prompt for a key press, the second item in C, that is, #A, is colored. The program prompts:

```
Press space bar to digress
```

You respond. The name C and the item number 2 are displayed at the bottom of the stack to show that a digression has been made from item

2 of list C. List C is deemphasized, and list A is highlighted. You are prompted to press the space bar again. You respond, and the first item in list A, the word "dog," is colored and copied into the output line. The program prompts for another key press for the next item. Successive key presses move the coloring through the successive items of list A and copy the items onto the output line. After the final item, the word "book," has been copied from list A, the program prompts:

```
Press space bar to revert
```

You respond. List A is deemphasized. List C is highlighted again, and the second item turns blue to show it is the item that caused a digression which has ended. The reference to this item is erased from the stack. You are prompted to press the space bar for the next item. You respond. The third item in list C, the word "cat," is colored and copied into the output line. The output line now contains the partially expanded list: keys, dog, chair, pipe, rug, cup, book, cat.

The next key press moves the coloring to the fourth item of C, the word "bird," and copies it onto the output line. The next key press moves the coloring to the fifth item of C; that is to #B, the reference to list B. You are prompted to press the space bar to digress. You respond. The list name C and the number 5, which identifies the current item in C, appear at the bottom of the stack on the screen.

List C is deemphasized and list B is highlighted. The program now works through the successive items of list B, colors these in turn, and copies them onto the output line. After the final item, the word "shoes," has been copied, the program prompts again to press the space bar to revert. You respond. List B is deemphasized. List C is highlighted again. The fifth item in list C, that is, #B, turns blue and the stack clears. The program recognizes that the end of the main list has been reached and prompts for a key press to end the overall process. You respond, and the prompt to keep the current list or return to the main menu reappears.

If you type K and then enter F for the main list of preset data, you can single-cycle the program through an unraveling that involves digressions from digressions several times over. If you press I in response to the main menu, the program prompts for a set of up to 10 lists with up to 10 items in each list. You can then trace the expansion of the lists and watch the stack wax as the program digresses from one list to another and wane when it reverts to the point from which a digression was made. Because the output list often will not fit on a single line of the screen, it is "scrolled" leftward when it exceeds 80 characters.

The basic principle demonstrated by the program is very simple:

pushing data about the point from which a digression is made and popping this data to revert when necessary. Despite its simplicity, however, the principle is of immense importance in innumerable situations.

Program Variables

IC	The maximum number of lists
JC	The maximum number of items in each list
A$(i)	The name of the ith list
JLIM(i)	The number of items in the ith list
INLIST$(i,j)	The jth item in the ith list
B$(i)	The ith list stored as a single string for display
SC	The capacity of the stack
STACK(k,1)	The number that identifies the list from which a digression was made
STACK(k,2)	The item number in the list that caused the digression
STACK(k,3)	The character position of the start of the item in the list treated as a string
VCSB	The position of the base of the stack on the screen
HCS	The position of the left edge of the stack
BAR$	The graphics character that forms the sides of the stack
HZL$	The graphics character that forms the base of the stack
CSR$	The code that moves the cursor one position to the right
S$	A string of 90 spaces used in PRINT statements to force double spacing
MS$	The string that is displayed in the current prompt or diagnostic
I	The subscript of the list under current attention in the array INLIST$
J	The subscript of the item under current attention in that list
V	The index of the loop that draws the sides of the stack
H$	The name of the head list to be expanded
TOP	The subscript, in the array STACK$, of the data about the item currently at the top of the stack
D$	The name of the list under current attention
ENDED	The switch that is turned on when the head list has been fully expanded
C$	The portion of the expanded list currently displayed on the screen

DIGRESS	The switch that is turned off when the program starts to work through a list and is turned on when it encounters a reference to a subsidiary list
A$	The item under current attention
HC	The horizontal cursor position of that item
T	The subscript of entries in the array STACK in the loop that checks for cyclic definition
Z$	The one-byte response to prompts for a key press
Q	The position of the next delimiter (colon or comma)
U$	The string that is typed in response to the prompt for a list when data is being input and then the residue of this string when successive elements are transferred to INLIST$
W$	The portion of this string that consists of an element of the list before any spaces have been removed from the beginning and end

Program Notes

The expansion of a list begins by (1) setting TOP to 0 to nullify the stack, (2) finding the subscript I of the head list, and (3) turning the switch ENDED to the off value. An outer WHILE . . . WEND loop then cycles until ENDED shows that the expansion is complete. To explain how this works, we first consider the behavior when all the items in the head list go directly into the output. The DIGRESS switch is set to the off value. The subscript J is set to 1. An inner WHILE . . . WEND loop works through the list without encountering a reference to another list. The DIGRESS switch remains off. No data is pushed onto the stack. Consequently TOP is still 0, and the ENDED switch is turned on and cycling ends.

Now suppose that an item in the head list refers to another list. A # is found at the beginning of the item. The DIGRESS switch is turned on. D$ is set to the name of the sublist. The current values of I and J are pushed onto the stack by the subroutine at line 5000. I is reset to point to the sublist. A new cycle through the outer WHILE . . . WEND loop is started to work through the sublist.

The new cycle through the outer loop turns the DIGRESS switch off and resets J to 1. The inner WHILE . . . WEND loop begins to operate in the manner already described. Suppose that all the items in the sublist go directly into the output. The inner loop is repeated for successive elements of the sublist until the end is reached. The statements within the loop are executed in the same manner as in the progression through the head list in the first example. On leaving the loop, however, TOP is 1 this time because data was pushed onto the stack when the program found a digression to be necessary. As a consequence,

```
1ØØ REM * INCLIST expands a list containing references to sublists *
1Ø1 :
11Ø  ON ERROR GOTO 9ØØØ                                    'set error trap
12Ø  KEY OFF: SCREEN Ø                                     'screen set up
13Ø    WIDTH 8Ø: COLOR 7,Ø
14Ø  CLS
15Ø  IC=1Ø                                          'limits on # of lists
16Ø  JC=1Ø                                          'and items per list
17Ø  DIM A$(IC), JLIM(IC), INLIST$(IC,JC), B$(IC)     'space for lists
18Ø  SC=2Ø                                              'stack capacity
19Ø  DIM STACK(SC,3)                                    'stack capacity
2ØØ  S$=SPACE$(9Ø)                          'string to force double spacing
21Ø  BAR$=CHR$(179)                              'vertical bar character
22Ø  HZL$=STRING$(4,196)                           'four horizontal bars
23Ø  CSR$=CHR$(28)                                  'cursor right code
24Ø  VCSB=22                          'vertical position of base of stack
25Ø  HCS=61                               'horizontal position of stack
299 :
3ØØ REM * main prompt *
31Ø  MS$="Type P for preset data, I to input data, Q to quit:"
32Ø  GOSUB 91ØØ
33Ø  ON INT((1+INSTR("PpIiQq",R$))/2) GOTO 4ØØ,8ØØØ,1ØØØØ          'branch
399 :
4ØØ REM * transfer preset data *
41Ø  MS$="Please wait - setting up data"
42Ø  GOSUB 95ØØ                                        'display the message
43Ø  READ ILIM                                          '# of lists
44Ø  FOR I=1 TO ILIM
45Ø   READ A$(I), JLIM(I)                           'name & # of items
46Ø   FOR J=1 TO JLIM(I)
47Ø    READ INLIST$(I,J)                             'transfer an item
48Ø  NEXT J,I
49Ø  RESTORE
599 :
6ØØ REM * initial display for a list expansion *
61Ø  CLS
62Ø  FOR I=1 TO ILIM
63Ø   GOSUB 2ØØØ                            'form display string from ith list
64Ø   GOSUB 3ØØØ                                          'display it
65Ø  NEXT I
66Ø  FOR V=VCSB-SC-1 TO VCSB-1
67Ø   LOCATE V,HCS
68Ø    PRINT BAR$+"   "+BAR$+"   "+BAR$           'draw sides of stack
69Ø  NEXT V
7ØØ  LOCATE VCSB,HCS                                    'draw the base
71Ø  PRINT CHR$(192)+HZL$+CHR$(193)+HZL$+CHR$(217)
72Ø  MS$="Name of main list: "                      'prompt for name
73Ø  GOSUB 95ØØ                                        'of main list
74Ø  INPUT; "",H$
999 :
1ØØØ REM * main cycle *
1Ø1Ø  TOP=Ø                                           'initialize the stack
1Ø2Ø  D$=H$
1Ø3Ø  GOSUB 25ØØ                                        'find the main list
1Ø4Ø  ENDED=Ø                                        'termination switch off
1Ø5Ø  C$=""                                          'nullify output string
1199 :
12ØØ  WHILE NOT ENDED                               'loop until end condition
121Ø  DIGRESS=Ø                                     'digression switch off
122Ø  J=1                                           'item # in current list
```

```
1230    HC=7                                           'cursor for highlighted item
1240    MS$="Press space bar for 1st item"                      'prompt
1250      GOSUB 9200
1399  :
1400    WHILE J<=JLIM(I) AND NOT DIGRESS               'loop through items
1410      COLOR 12
1420      GOSUB 3500                                   'highlight jth item
1430      A$=INLIST$(I,J)
1440      IF LEFT$(A$,1)="#" THEN 1500         'jump if it is a cross reference
1450      IF C$="" THEN C$=A$ ELSE C$=RIGHT$(C$+", "+A$,79)      'put into output
1460      LOCATE 25,1                                         'and display
1470      PRINT C$;
1480      GOSUB 4000                             'move attention to next item
1490      GOTO 1570                                   'and to end of loop
1500      DIGRESS=-1                             'cross reference found
1510      D$=MID$(A$,2)                                 'name of sublist
1520      MS$="Press space bar to digress"                    'prompt
1530      GOSUB 9200
1540      GOSUB 3000                             'dehighlight current list
1550      GOSUB 5000                       'push data about it onto stack
1560      GOSUB 2500                       'start the referenced sublist
1570    WEND
1699  :
1700    IF DIGRESS THEN 1830                       'jump if digression needed
1710      IF TOP<1 THEN ENDED=-1                   'test for termination
1720    MS$="Press space bar to "+MID$("revert end",1-7*ENDED,7)
1730      GOSUB 9200                              'prompt for key press
1740      GOSUB 3000                              'dehighlight the list
1750      IF ENDED THEN 1830                   'jump if it was the head list
1760        GOSUB 6000                                 'pop the stack
1770        COLOR 15
1780        GOSUB 3000                       'highlight the reentered list
1790        COLOR 3
1800        GOSUB 3500                       'color the reentered item
1810        GOSUB 4000                       'move on to next item
1820        GOTO 1400
1830    WEND
1840    IF POS(0)=1 THEN 1900
1899  :
1900    GOSUB 3000                               'dehighlight the head list
1910    MS$="Type K to keep present lists, M for main menu"
1920      GOSUB 9100
1930    CLS
1940    IF R$="M" OR R$="m" THEN 300 ELSE 600      'main menu or keep lists
1999  :
2000 REM * convert a list into a string $
2010    B$(I)=LEFT$(A$(I),4)+": "                'start ith display string
2020    IF LEN(B$(I))<6 THEN B$(I)=B$(I)+SPACE$(6-LEN(B$(I)))   'space if needed
2030    FOR JP=1 TO JLIM(I)
2040      B$(I)=B$(I)+INLIST$(I,JP)                'concatenate jth item
2045      IF JP<JLIM(I) THEN B$(I)=B$(I)+", "
2050    NEXT JP
2060    B$(I)=LEFT$(B$(I),60)                       'truncate if necessary
2070    RETURN
2499  :
2500 REM * start a (sub)list *
2510    FOR I=1 TO ILIM                          'search for name of list
2520      IF D$=A$(I) THEN 2570
2530    NEXT I
2540    I=0                                         'invalid name
```

```
2550    MS$="Invalid (sub)list name - press space bar"
2555    GOSUB 9200
2560    IF TOP=0 AND C$="" THEN GOTO 720 ELSE ERROR 250    'retry or error trap
2570   FOR T=1 TO TOP                              'check for cyclic definition
2580    IF I=STACK(T,1) THEN 2630
2590   NEXT T
2600   COLOR 15                                       'highlight the new list
2610    GOSUB 3000
2620   RETURN
2630   MS$="Cyclic definition"                      'cyclic definition found
2640    ERROR 250                                        'trigger error trap
2999 :
3000 REM * display a (sub)list *
3010   LOCATE 2*I-1,1
3020   PRINT B$(I)                                 'display string form of list
3030   COLOR 7                                              'default color
3040   RETURN
3499 :
3500 REM * display an item *
3510   LOCATE 2*I-1,HC
3520    PRINT INLIST$(I,J);                       'display jth item of ith list
3525    IF J<JLIM(I) THEN PRINT ",";
3530    COLOR 7                                              'default color
3540    RETURN
3999 :
4000 REM * get next item *
4010   IF J=JLIM(I) THEN 4060                   'bypass prompt if final item
4020    MS$="Press space bar for next item"
4030    GOSUB 9200                                      'prompt for next item
4040    COLOR 15                                      'de-color current item
4050    GOSUB 3500
4060    J=J+1                                            'increase the cursor
4070    HC=POS(0)+1                                           'text cursor
4080    RETURN
4999 :
5000 REM * push data onto the stack *
5010   IF TOP<SC THEN 5040                          'test for stack overflow
5020    MS$="stack overflow"
5030    ERROR 250                                        'trigger error trap
5040   TOP=TOP+1                                   'increase pointer to top
5050    STACK(TOP,1)=I                                  'stack the list #
5060    STACK(TOP,2)=J                                      'the item #
5070    STACK(TOP,3)=HC                                  'and the cursor
5080    LOCATE VCSB-TOP,HCS+1
5090    PRINT USING "\    \!###"; A$(I),CSR$,J        'display stacked data
5100    RETURN
5999 :
6000 REM * pop data from the stack *
6010   IF TOP>0 THEN 6040                     'jump unless stack is empty
6020    MS$="Tried to pop when empty"
6030    ERROR 250                                        'trigger error trap
6040    I=STACK(TOP,1)                                     'pop list #
6050    J=STACK(TOP,2)                                        'item #
6060    HC=STACK(TOP,3)                                  'and cursor
6070   LOCATE VCSB-TOP,HCS+1
6080    PRINT SPC(4) CSR$ SPC(3)                     'erase data from stack
6090    TOP=TOP-1                                   'decrease pointer to top
6100   RETURN
6999 :
7000 REM * preset data *
```

```
7010    DATA 6                                          :'length of preset list
7020    DATA A, 6, dog, chair, pipe, rug, cup, book
7030    DATA B, 5, hat, coat, tie, scarf, shoes
7040    DATA C, 5, keys, #A, cat, bird, #B
7050    DATA D, 5, tree, car, #A, bus, #C
7060    DATA E, 5, train, plane, #A, #C, table
7070    DATA F, 6, #A, #C, boat, lamp, #E, #A
7999    :
8000 REM * input data *
8010    CLS                                            'instructions for typing data
8020    LOCATE 5,11
8030    PRINT      "Start each list on a new line. Limit each list to 10 items"
8040    PRINT S$ "and 60 characters. Start each list with its name followed"
8050    PRINT S$ "a colon. Leave a comma between each item and its successor."
8060    PRINT S$ "In a list, put a # before the name of another list, as in"
8070    PRINT S$ "          clump: dog, pipe, #lump, chair"
8080    PRINT S$ "Press the Enter key at the end of each list, and an extra"
8090    PRINT S$ "time after the final list. Do not type more than 10 lists."
8100    PRINT
8110    I=0
8199    :
8200    LINE INPUT U$                                       'input a list
8210    IF U$=SPACE$(LEN(U$)) THEN 8770                     'jump if null
8220    IF I<IC THEN 8300                              'jump if within limit
8230    PRINT S$ "Limit has been reached on number of lists"
8240    PRINT S$ "Type E to expand a list, S to start again"
8250    R$=INPUT$(1)
8260     IF R$="S" OR R$="s" THEN 8000 ELSE 8770             'remedial action
8299    :
8300    I=I+1                                              'increase list #
8310    Q=INSTR(U$,":")                                    'find colon
8320    IF Q>0 THEN 8350
8330    MS$="No colon"                                      'error condition
8340    GOTO 8720
8350    W$=LEFT$(U$,Q-1)                                    'extract left
8360    U$=MID$(U$,Q+1)                                   'and right substrings
8370    GOSUB 8800                                     'remove outside spaces
8380    A$(I)=W$                                             'list name
8390    J=0
8499    :
8500    WHILE LEN(U$)>0 AND J<JC                        'loop through items
8510     Q=INSTR(U$,",")                                    'find space
8520     IF Q=1 THEN U$=MID$(U$,2): GOTO 8590               'leading space
8530     IF Q>0 THEN W$=LEFT$(U$,Q-1) ELSE W$=U$        'allow for final item
8540     GOSUB 8800                                    'remove outside spaces
8550     IF LEN(W$)=0 THEN 8590                             'jump if null
8560     J=J+1                                             'increase item #
8570     INLIST$(I,J)=W$                                   'store the item
8580     IF Q>0 THEN U$=MID$(U$,Q+1) ELSE U$=""         'residual string
8590    WEND
8600    IF U$<>SPACE$(LEN(U$)) THEN 8700               'jump if too many items
8610    JLIM(I)=J                                          'set item count
8620    GOTO 8200                                      'go back for another list
8699    :
8700    PRINT "Too many items in a list - type T to truncate, S to start again:"
8710     Z$=INPUT$(1)                                  'wait for key press
8720     IF Z$="T" OR Z$="t" THEN 8610
8740     I=I-1
8750     PRINT
8760     GOTO 8200                                                 'try again
```

```
8770    ILIM=I                                                  '# of lists
8780    GOTO 600                                    'go to display and expand action
8799 :
8800 REM * remove leading spaces *
8810    WHILE LEFT$(W$,1)=" "
8820     W$=MID$(W$,2)                                           'remove left space
8830    WEND
8840    WHILE RIGHT$(W$,1)=" "
8850     W$=LEFT$(W$,LEN(W$)-1)                                 'remove right space
8860    WEND
8870    RETURN
8999 :
9000 REM * error trap *
9010    IF ERR<>250 THEN MS$="Error"+STR$(ERR)+" in line"+STR$(ERL) 'system error
9020     RESUME 9030
9030    MS$=MS$+" - press space bar"                                  'diagnostic
9040     GOSUB 9200
9050     GOTO 300                                        'go back to main menu
9099 :
9100 REM * prompt for an option *
9110    GOSUB 9500
9120    R$=INPUT$(1)
9130    RETURN
9199 :
9200 REM * prompt for a single key press *
9210    GOSUB 9500
9220     Z$=INPUT$(1)                                          'wait for key press
9230    RETURN
9499 :
9500 REM * prompt for a datum *
9510    LOCATE 23,1
9520    PRINT TAB(79)                                         'clear message line
9530    LOCATE 23,1
9540    PRINT MS$;                                               'display prompt
9550    RETURN
9999 :
10000 REM * quit *
10010    MS$=""
10020    GOSUB 9500                                            'clear message line
10030    COLOR 7                                                 'default color
10040    ON ERROR GOTO 0                                       'error trap address
10050    END                                                          'quit
```

ENDED is not reset by statement 1710. As a result the subroutine at line 6000 pops the stack.

This action resets I and J to point to the item in the main list that caused the digression. Then the subroutine at line 4000 moves attention to the next item in the Ith list (or finds that the final item of the list has been reached). Control goes back to the start of the inner loop to continue working through the list to which it has returned. This continuation begins with the item currently identified by J, that is, the item following the one that caused the digression. If the digression was caused by the final item in the head list, the inner loop is bypassed because the WHILE condition J<=JLIM(I) is not true.

In general, the flow of control is a simple extension of the patterns that have been described. Whenever the statements in the inner loop find the name of another list in the list under current attention, a digression is started by turning the DIGRESS switch on. The data about the item that is causing the digression is pushed onto the stack. The inner loop is left, and a new cycle through the outer loop starts to process the new list. The inner loop continues the processing of the new list. Whenever the inner loop reaches the end of a list, the ensuing statements in the outer loop pop the stack. The inner loop is restarted from the point after the item that caused the digression which has just been completed. Finally, the head list is ended. The ENDED switch is turned on, and cycling through the outer loop terminates.

Suggested Activities

1. Expand INCLIST to allow the inclusion of a repetition factor before a reference to a subsidiary list. Thus, the definition A: 2#B, #C, 3#D would be equivalent to A: #B, #B, #C, #D, #D, #D.

2. Write a program that (a) prompts for a positive integer n and (b) generates all n-character strings that consist of the letters A to E. Use a stack.

3. Expand Activity 2 to use different sets of letters in the successive positions of the output strings. Let the user specify these in response to prompts.

4. Write a program A that (a) prompts for a positive integer n, (b) constructs a program B to generate all n-character strings that consist of the letters A to E by means of a nest of n FOR . . . NEXT loops, (c) writes program B onto disk, and (d) executes it. Use a stack in program A.

4.3 Computing a Polish Expression

The instruction "take the numbers three and five and add them" is a perfectly natural way to describe what should be done to evaluate the expression $5 + 3$. The Polish mathematician Lukasiewicz developed a whole system for writing arithmetic that puts each symbol, such as $+$, $-$, $\times$, or $/$, after the pair of numbers it combines instead of between them. The expression $5 + 3$ is written 5 3 $+$ in this *reverse Polish notation,* as it is called. Correspondingly, $5 - 3$ is written 5 3 $-$, and 5×3 is written 5 3 $\times$. Lukasiewicz developed the notation to avoid the need for parentheses in certain kinds of work, and his way of representing expressions has turned out to be extremely important for the application of computers to arithmetic and other problems.

The advantage of using Polish expressions can be seen by looking at the ordinary expression $(3 + (5 - 4) \times (7 - 2))$. To evaluate it, you have to use the $\times$ symbol after doing the two subtractions, and only then can you use the $+$ symbol. Thus you have to dodge forward and backward to deal with the mathematical symbols. The corresponding Polish expression is

3 5 4 − 7 2 − × +

You evaluate this by starting at the left and going forward until you reach an arithmetic symbol. You use the symbol to combine the two numbers on its left and write the result in place of them. At this point you have the numbers 3 1 in reserve (the 1 came from subtracting 4 from 5), and you go on to the 7, the 2 and then to the second minus symbol. You use this symbol to combine the two numbers on its left, that is, the 7 and 2, to get 5. Now you have the numbers 3 1 5 in reserve. The next symbol is ×. You use it to multiply the two numbers that precede it at this time. These are 1 and 5, giving 5, so the two numbers 3 and 5 are left in reserve. Now you encounter the + symbol and use it to add the 3 and 5, which gives you the final answer, 8.

Expressions like 5 3 + and 3 5 4 − 7 2 − × + are also called *suffix expressions* and *postfix expressions,* and the term "infix" is used for expressions, like $5 + 3$, that contain the symbols between the things they combine. A symbol, such as $+$, $-$, $\times$, or $/$, that defines the result of combining two numbers is called a *binary operator*. The numbers that it combines are called *operands*.

The evaluation process that has been described is programmed by:

1. Working through the expression from left to right

2. Pushing each number onto a stack when it is encountered

3. Popping the top two numbers currently at the top of the stack and combining them when a binary operator is encountered

4. Pushing the result of a binary operation back onto the stack

The program VALPOST begins by prompting:

```
    Type a postfix expression
       containing one-letter
    variables and the operators
          +  -  *  /   ^
    for example: a b + c d - * e /
    and then press the Enter key

type here:
```

The symbols * and ^ used in the prompt are the input for multiplication and exponentiation, respectively. To make a simple start, enter
the expression uv+ that is even shorter than the example in the
prompt. Another message appears lower on the screen.

```
Enter integers between -99 and 99
for the variables when prompted.
```

The values that you type are restricted because of the way the output is
displayed. The program prompts for the values of u and v. Suppose you
enter the numbers 7 and 12. The screen clears and the input expression
is redisplayed at the top, followed by the values of u and v. An empty
stack appears at the right of the screen, and you are prompted to press
the space bar.

```
Original expression: u v +

u=7     v=12

Press space bar
to proceed
```

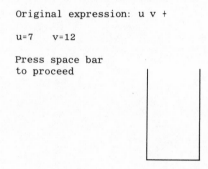

Press the space bar. Attention is directed to the first character in the
input expression, the variable u. It turns red. The program displays a
message that u is a variable and prompts you to press the space bar to
push its value onto the stack. You respond. The value, 7, appears at the
bottom of the stack.

The prompt disappears. Attention moves to the next character in the
original expression, shown by its turning red. The program displays a
message that v is a variable and prompts you to press the space bar to

push its value onto the stack. The v reverts to its original color; its value 12 appears above the 7 in the stack; and the message area clears.

```
|           |
|    12     |
|    7      |
|_____|
```

Attention moves on to the next character in the expression. This is the + symbol. The program displays a message that + is an operator and prompts you to press the space bar to pop the top number. You respond, thereby making the 12 disappear from the stack and appear alongside the message. The program prompts you to press the space bar again. You do so, and the only number left in the stack, the 7, disappears from the stack and reappears in the message area. You are prompted to press the space bar to combine the values. You respond, and the display becomes:

```
Original expression: u v +

u=7    v=12

+ is an operator

Press space bar
     to pop top
        number
                    12
        and again
                    7
Press space bar
     to combine
                    19

and to push
the result
onto stack
```

You respond to the prompt at the bottom of the display. The message area clears, and the 19 appears at the bottom of the stack.

```
|           |
|           |
|    19     |
|_____|
```

The entire expression has been processed. As soon as you press the space bar again, the message

```
Final result is 19

(N)ew values,  (E)xpression,  or  (Q)uit.
```

appears. When the expression is more complicated, data rises and falls in the stack to a greater extent. For example, the contents of the stacks (here drawn narrower than on the screen to save space) at successive stages of the evaluation of:

```
a  b  c  -  d  e  -  *  +
```

when

```
a=3,  b=5,  c=4,  d=7,  e=2
```

are

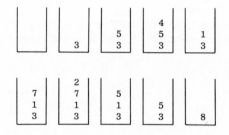

The character at the top of the stack is displayed in red for emphasis. Diagnostic messages are displayed

1. If you type an input expression that is more than 19 characters long
2. If the expression contains any characters besides spaces (which are ignored), letters, and the symbols +, −, *, /, and ^
3. If the expression is not valid, for example, if it is a + * or abc +
4. If overflow occurs or division by zero is attempted
5. If errors occur of other kinds that were not anticipated

The stack is wide enough for the largest possible integers or floating point numbers that can occur. The following "snapshot" of an interim stage of the evaluation of a rather large result shows how the alignment prevents long numbers from encroaching on the displayed stack.

```
Original expression:

a b c d e f g h i j * * * * * * * * *

a=99   b=98   c=97   d=96   e=95   f=94
g=93   h=92   i=91   j=90

* is an operator

Press space bar
      to pop top
           number
                  753480
         and again
                        93
Press space bar                  94
      to combine                 95
          7.007364E+07           96
                                 97
and to push                      98
the result                       99
onto stack
```

The following snapshot shows an error message in a circumstance with which the system cannot cope: raising 99 to the power 98 causes numerical overflow.

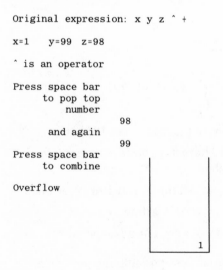

```
Original expression: x y z ^ +

x=1    y=99  z=98

^ is an operator

Press space bar
      to pop top
           number
                        98
         and again
                        99
Press space bar
      to combine

Overflow
```

```
(N)ew values, (E)xpression, or (Q)uit:
```

The structure of the program is quite simple. Only a few statements are needed to evaluate a postfix expression and to check its validity. Most of the program formats the display and provides the running commentary.

Program Variables

T$(k)	The kth non-blank character in the input expression
ALPHABET$	The string consisting of the lowercase letters a to z and the capitals A to Z
DIGITS$	The string consisting of the digits 0 to 9
OPS$	The string consisting of the operator symbols +, −, *, /, and ^
STC	The horizontal cursor position of data in the stack
W	The width of the stack
U$	A string of 40 spaces used to force double spacing in some PRINT statements
S$	The input expression to be evaluated
IMAX	The number of characters in it
I	The index of the non-blank characters in loops that deal with the expression
C$	The character under current attention in some of the loops
VO	The vertical cursor position at which the expression is redisplayed
HO	The horizontal position of the character under current attention
VN	The vertical cursor position of the first line of the area in which prompts are displayed
P	The number that identifies a prompt internally
PLIM	The total number of prompts
VBLS$	The string that consists of the names of variables for which values have been provided
D	The index of a delay loop
J	The index of the variables in the order of their first occurrence in the input expression
V(j)	The value of the jth variable
K	The numbers 1 to 5 for the operators +, −, *, /, ^, respectively
NOUT	The number just popped from the stack

V1,V2	The two numbers that were just popped and that are being combined
V	The result
R$	The one-character response to the prompt for further action
STACK(k)	The kth number currently in the stack, counting from the number at the bottom of the stack
TOP	The number of items currently in the stack, that is, the index of the top item in STACK
L	The line number in the loop that clears the message area
FN DS$(x)	The string representation of x without leading spaces

Program Notes

After some housekeeping, a FOR . . . NEXT loop cycles through the successive characters of the original expression. When a variable is encountered for the first or only time in this loop, the fact is shown by (1) the presence of the character in the string ALPHABET$ and (2) the absence of the character from the string VBLS$.

```
770 IF INSTR(ALPHABET$,T$(I))=0 OR INSTR(VBLS$,T$(I))>0 THEN 860
```

When a new variable is encountered, the program prompts for its value. The name of the current variable is appended to the string VBLS$ when the value has been specified.

```
850 VBLS$=VBLS$+T$(I)
```

The evaluation algorithm is embodied in a FOR . . . NEXT loop

```
2210 WHILE I<=IMAX
...
3150 WEND
```

This loop progresses through the successive characters of the input expression. The INSTR function is used to identify the character in the string of variables VBLS$ or in the string of operators OPS$. The appropriate value is pushed when the character is a variable. The two characters at the top of the stack are popped and combined appropriately when the character under current attention in the expression is an operator.

Subroutines are used to push and to pop the stack. The push sub-routine

1. Increases the value of TOP by 1
2. Makes V the new TOPth element of STACK
3. Displays V in the appropriate position

```
4020 TOP=TOP+1
4030 STACK(TOP)=V
4040  LOCATE 22-TOP, STC
4050   PRINT TAB(39-LEN(STR$(V)))  V
```

The pop subroutine

1. Checks that the stack is not empty
2. Sets NOUT to the top element
3. Erases NOUT from the screen
4. Decreases TOP by 1

```
5010 IF TOP<1 THEN 8500
5020  NOUT=STACK(TOP)
5030  LOCATE 22-TOP, STC
5040   PRINT SPC (W)
5050 TOP=TOP-1
```

Error conditions are recognized when an attempt is made to pop an empty stack and when the loop ends before the stack is empty.

Suggested Activities

1. Logical variables can have the value true (T) or false (F) and can be combined by the binary logical operators AND, denoted by &, and OR, denoted by |. AND operators take precedence over OR, just as multiplication of numerical variables takes precedence over addi-tion. Thus, if p, q, and r are logical variables, p&q|r is equivalent to (p&q)|r. Adapt the program VALPOST to evaluate a logical expres-sion that is provided in postfix form, for example, pq&r|. Let the user type T or F in response to the prompts for the values of the indi-vidual variables.

2. Suppose (a) that you have a collection of objects, (b) that you have defined several different (but possibly overlapping) sets of those objects, and (c) that you are interested in the membership of further sets that are formed by taking the union, intersection, and differ-

```
100 REM * VALPOST evaluates postfix expression *
101 :
110  ON ERROR GOTO 9000                                         'set error trap
120   KEY OFF: SCREEN 1: COLOR 1,0                              'screen set up
130  DEF FN DS$(M)=MID$("-",2+SGN(M),1)+MID$(STR$(M),2)          'to align #s
140  DIM T$(40)                                       'elements of expression
150  OPS$="+-*/^"                                      'reference strings
160   DIGITS$="0123456789"
170   ALPHABET$="abcdefghijklmnopqrstuvwxyzABCDEFGHIJKLMNOPQRSTUVWXYZ"
180  W=16                                                       'width &
190  STC=24                                            'position of stack
200  U$=SPACE$(40)                                     'formatting string
210  GOSUB 6000                                       'set up the prompts
299 :
300 REM * prompt for an expression *
310  CLS                                              'prompt for an expression
320   PRINT    "        Type a postfix expression"
330   PRINT U$ "         containing one-letter"
340   PRINT U$ "        variables and the operators"
350   PRINT U$ "           +  -  *  /   ^"
360   PRINT U$ "    for example:";
370  POKE &H4E,1                                               'green
380    PRINT " a b + c d - * e /"
390  POKE &H4E,3                                               'yellow
400   PRINT U$ "      and then press the Enter key"
410   PRINT U$ " type here: ";
420   LINE INPUT S$                                  'store the expression
499 :
500 REM * separate the non blank characters *
510   IMAX=0
520  FOR I=1 TO LEN(S$)                              'loop through characters
530   C$=MID$(S$,I,1)                                 'ith character
540   IF C$ = " " THEN 580                            'jump if space
550   IF INSTR(ALPHABET$+OPS$,C$)=0 THEN GOTO 8500    'jump if invalid
560   IMAX=IMAX+1
570   T$(IMAX)=C$                                             'concatenate
580  NEXT I
590   IF IMAX < 20 THEN 710
600   PRINT U$ " Please use less than 19 letters and"   'prompt for retry
610   PRINT U$ " symbols. Press any key to try again.";
620   GOSUB 7500                                     'wait for key press
630    GOTO 300
699 :
700 REM * specify values of the variables *
710   PRINT U$ "    Enter integers between -99 and 99"   'prompt for values
720   PRINT U$ "    for the variables when prompted"
730   PRINT
740  J=0                                              'initialize count and
750  VBLS$=""                                         'string of variables
760  FOR I=1 TO IMAX
770  IF INSTR(ALPHABET$,T$(I))=0 OR INSTR(VBLS$,T$(I))>0 THEN 860
780   J=J+1                                                   'new one
790   LOCATE CSRLIN,15
800    PRINT T$(I) "=";                                'prompt for value
810    INPUT "", V(J)
820    IF ABS(V(J))<100 AND V(J)=INT (V(J)) THEN 850   'check validity
830     PRINT U$ "Please use an integer between -99 and 99"     'diagnostic
840     GOTO 790
850    VBLS$=VBLS$+T$(I)                     'append new variable name to string
860  NEXT I
```

```
 999 :
1000 REM * redisplay the original expression *
1010  CLS                                                      'redisplay
1020   PRINT "Original expression: ";
1030   IF IMAX>9 THEN PRINT U$ TAB(20-IMAX)             'align or center
1040   VO=CSRLIN                                            'vertical and
1050   HO=POS(0)                                       'horizontal cursor
1060    FOR I=1 TO IMAX
1070     PRINT T$(I)+" ";                         'display the ith character
1080    NEXT I
1090   LOCATE CSRLIN+2,1
1100   FOR JJ=1 TO J                                            'variables
1110    PRINT MID$(VBLS$,JJ,1) "=" FN DS$(V(JJ)) TAB(1+6*(JJ MOD 6))
1120   NEXT JJ
1130   VN=CSRLIN                                'vertical reference for prompts
1499 :
1500 REM * draw the stack *
1510   LINE (8*STC-12,72)-STEP (0,100)                 'left side of stack
1520   LINE -STEP(8*W+8,0)                                          'base
1530   LINE -STEP(0,-100)                                     'right side
1999 :
2000 REM * evaluate the expression *
2010   LOCATE VN+2,1
2020   P=1: GOSUB 7000                        'prompt to start evaluation
2030   GOSUB 8000                                   'clear message area
2060   I=1                                        'pointer to expression
2070   TOP=0                               'initialize pointer to stack
2199 :
2200 REM * loop through the characters *
2210   WHILE I<=IMAX                             'loop through characters
2220   C$=T$(I)                                          'next character
2230   POKE &H4E,2
2240    LOCATE VO,HO+2*I-2                 'turn it red in original expression
2250    PRINT C$
2260   LOCATE VN+2,1                                  'start the comment about
2270   PRINT C$;                                      'how it will be treated
2280   POKE &H4E,3
2290   J=INSTR(VBLS$,T$(I))                               'id # of variable
2300    IF J=0 THEN 2510                                 'jump if operator
2310   P=2: GOSUB 7000                    'prompt to push variable onto stack
2320    GOSUB 8000
2330   V=V(J)
2340   LOCATE VO,HO+2*I-2              'dehighlight in original expression
2350    PRINT C$
2360    GOSUB 4000                                  'push it onto stac
2410   GOTO 3140
2499 :
2500 REM * deal with an operator *
2510   K=INSTR("+-*/^",T$(I))                            'id # of operator
2520    IF K=0 THEN 8500                        'jump if not an operator
2530   P=3: GOSUB 7000                                 'prompt to pop
2540    GOSUB 5000                               'pop item from stack
2590    V2=NOUT
2600    LOCATE VN+7,21-LEN(STR$(V2))                      'display it in
2610    PRINT V2                                        'message area
2615   LOCATE VN+8,1                            .
2620   P=4: GOSUB 7000                             'prompt to pop again
2630    GOSUB 5000                               'pop another item
2650    V1=NOUT
2660    LOCATE VN+9,21-LEN(STR$(V1))                     'display it in
```

```
2670      PRINT V1                                                  'message area
2680      P=5: GOSUB 7000                                           'prompt to combine
2690      LOCATE VN+13,1                                            'for error message
2999 :
3000 REM * form the result and push it onto the stack *
3010   IF K=1 THEN V=V1+V2                                          'add
3020   IF K=2 THEN V=V1-V2                                          'subtract
3030   IF K=3 THEN V=V1*V2                                          'multiply
3040   IF K=4 THEN V=INT(V1/V2)                                     'divide
3050   IF K=5 THEN V=V1^V2                                          'exponentiate
3060     IF CSRLIN>VN+13 THEN 3600                          'jump if error occurred
3070     LOCATE VN+12,21-LEN(STR$(V))                           'display the result
3080     PRINT V
3090     P=6: GOSUB 7000                                 'prompt to push it onto stack
3100     GOSUB 4000                                                 'do so
3110     GOSUB 8000                                         'clear message area
3120     LOCATE VO,HO+2*I-2                                    'de-highlight the
3130     PRINT C$                                              'processed item
3140     I=I+1                                             'proceed to the next
3150   WEND
3499 :
3500 REM * display the final result *
3510   IF TOP<>1 THEN 8500                                      'end reached
3520   LOCATE VN+2,1
3530   P=7: GOSUB 7000                                          'prompt to pop
3540   GOSUB 8000
3550   GOSUB 5000                                             'pop the result
3560   LOCATE VN+2,1
3570   PRINT "Final result is" SPC(1+(NOUT>=0)) NOUT TAB(40)    'display it
3580   FOR D=1 TO 100: NEXT D                                        'delay
3600   LOCATE 24,1
3610   PRINT "(N)ew values, (E)xpression, or (Q)uit:" TAB(40)
3620   R$=INPUT$(1)
3630   ON (1+INSTR("NnEeQq",R$))\2 GOTO 3640,300,10000          'branch
3640   CLS                                                     'redisplay
3650   PRINT "Original expression: ";
3660   IF IMAX>9 THEN PRINT U$ TAB(20-IMAX)                    'align or center
3670     FOR I=1 TO IMAX
3680       PRINT T$(I)+" ";                             'display the ith character
3690     NEXT I
3700   GOTO 700
3999 :
4000 REM * push a value onto the stack *
4010   IF TOP>9 THEN 8500                               'error jump if stack is full
4020   TOP=TOP+1                                          'increase stack pointer
4030   STACK(TOP)=V                                         'store item in stack
4040     LOCATE 22-TOP,STC
4050     PRINT TAB(39-LEN(STR$(V))) V                       'display it in place
4060   RETURN
4999 :
5000 REM * pop the top number from the stack *
5010   IF TOP<1 THEN 8500                              'error jump if stack is empty
5020   NOUT=STACK(TOP)                                   'get the value of the item
5030     LOCATE 22-TOP,STC
5040     PRINT SPC(W)                                  'erase it from displayed stack
5050   TOP=TOP-1                                         'decrease stack pointer
5060   RETURN
5999 :
6000 REM * set up the prompts *
6010   PLIM=7                                                      '# of messages
```

```
6Ø2Ø   DIM PS$(PLIM,6)                              'array to hold them
6Ø3Ø   FOR P=1 TO PLIM                             'loop through messages
6Ø4Ø    READ PS$(P,Ø)                                   '# of lines
6Ø5Ø     FOR Q=1 TO VAL(PS$(P,Ø))                   'loop through lines
6Ø6Ø      READ PS$(P,Q)                              'transfer a line
6Ø7Ø     NEXT Q,P
6Ø8Ø   RETURN
65ØØ REM * text of prompts *
651Ø   DATA 2, "Press space bar", "to proceed"
652Ø   DATA 5, " is a variable", " ", "Press space bar", "to push its value",
              "onto the stack"
653Ø   DATA 5, " is an operator", " ", "Press space bar", "    to pop top",
              "      number"
654Ø   DATA 1, "     and again"
655Ø   DATA 2, "Press space bar", "     to combine"
656Ø   DATA 4, " ", "and to push", "the result", "onto stack"
657Ø   DATA 6, " ", "End reached", " ", "Press space", "bar to pop", "the stack"
6999 :
7ØØØ REM * display a prompt *
7Ø1Ø   FOR Q=1 TO VAL(PS$(P,Ø))
7Ø2Ø    PRINT PS$(P,Q);                             'display a prompt line
7Ø3Ø    IF CSRLIN<25 THEN PRINT                      'increase cursor
7Ø4Ø   NEXT Q
7Ø5Ø   GOSUB 75ØØ                                   'wait for key press
7Ø6Ø   RETURN
75ØØ REM * wait for key press *
752Ø   Z$=INPUT$(1)                                 'wait for key press
753Ø    IF R$=CHR$(27) THEN 1ØØØØ                    'quit if Esc
754Ø   RETURN
7999 :
8ØØØ REM * clear the message area *
8Ø1Ø   FOR L=VN+2 TO VN+17                          'clear the message area
8Ø2Ø    LOCATE L,1
8Ø3Ø     PRINT SPC(22)
8Ø4Ø   NEXT L
8Ø5Ø   RETURN
85ØØ REM * invalid input *
851Ø   LOCATE 24,1                                  'diagnostic for conditon
852Ø    PRINT "invalid expression - press space bar";   'recognized by program
853Ø    GOSUB 75ØØ
854Ø    GOTO 36ØØ
8999 :
9ØØØ REM * error trap *
9Ø1Ø   LOCATE 24,1                                  'system error
9Ø2Ø    PRINT "Error" ERR "in line" ERL "- press space bar";   'diagnostic
9Ø3Ø    GOSUB 75ØØ
9Ø4Ø    RESUME 1ØØØØ
9999 :
1ØØØØ REM * quit *
1ØØ1Ø   POKE &H4E,3                                 'default color
1ØØ2Ø   LOCATE 22,1                                 'for Ok message
1ØØ3Ø   ON ERROR GOTO Ø                             'reset error address
1ØØ4Ø   END
```

ences of sets which already have been defined. If the individual objects are numbered 1, 2, . . . , then a set of them can be represented by a string in which the kth character is 1 if the object is in the set and 0 if it is absent. Reserve the letters U, I, D, and S for the union, intersection, difference, and symmetric difference operators. Adapt VALPOST to evaluate set expressions that contain those operators.

3. An expression that contains binary operators can be written in prefix form by putting the operator before its operands. Thus, the prefix form of a + b*c is +a*bc and the prefix form of (a + b)*c is * + abc. Write a program to evaluate an infix expression. (*Hint:* To process a valid expression by a left-to-right scan, each operator is pushed onto the stack whatever is at the top. A number is pushed if an operator is at the top. When a number is under attention in the input and a number is at the top of the stack, the latter number and the operator immediately beneath it are popped and the two numbers are combined. The result is treated as if it were a number that had just been encountered in the input.)

4.4   Converting an Expression to Postfix

The program POSTCON converts infix expressions to the postfix form that VALPOST evaluates. POSTCON begins by prompting

```
Type a simple algebraic expression
that contains one-letter variables,
     and +, -, * and / symbols,
          for example
       a + b * c - d * p / q
 and then press the Enter key
 type here:
```

Suppose you respond by typing the very simple expression x + y and pressing the Enter key. The display changes to

```
Original expression: x + y
Postfix expression:
```

The prompt to press the space bar to go on appears at the bottom of the screen. The empty stack is drawn in the center portion of the screen. An @ symbol is displayed at the end of the original expression. Another @ appears at the bottom of the stack.

The infix-to-postfix conversion now progresses in response to successive key presses in the manner described in the next few paragraphs.
The program displays the message:

```
move forward
```

You press the space bar in response to this and to other messages that
appear later. Attention is focused on the variable x in the original
expression. The message changes to:

```
compare:    x  @
```

This shows that the x in the original expression will be compared with
the @ at the top of the stack. The "precedence values" of x and @ are
displayed under the prompt:

```
compare:    x  @
----------------
precedence: 3  0
```

The relative precedence of (1) the character under current attention in
the original expression and (2) the character at the top of the stack
determine whether (*a*) the character in the input expression is pushed
onto the stack or (*b*) the character at the top of the stack is popped into
the output expression. The precedence values that are used are arbitrary, except that their relative size must produce the correct outcome
in all the cases that can occur.

Accordingly, we use 3 as the precedence value of variables, 0 as the
precedence value of @, 1 as the precedence value of + and −, and 2 as
the precedence value of * and /. The program then shows that as the
result of the comparison, the x will be pushed onto the stack.

```
compare:    x  @
----------------
precedence: 3  0
            >
            push
```

The x is copied into the slot above the @ in the stack.

The message "move forward" reappears. Attention then moves on to the + sign in the original expression.

Throughout this activity, the item at the top of the stack and the element of the input expression under current attention are displayed in red to direct the viewer's attention to the center of activity. When an element of the input expression has been processed, it is deemphasized by another color change.

The comparison of the + in the input expression and the x at the top of the stack is displayed.

```
compare:    +  x
----------------
precedence: 1  3
               <=
             pop
```

Then this information is replaced by a check that the number of operators encountered so far is less than the number of variables. The check computes the "rank" of the expression that is produced by the popping. The rank is initialized to zero when the conversion begins. The rank of the popped item is added to the rank of the expression whenever popping occurs. The rank of an operator is -1, and the rank of a variable is 1.

```
popping - check ranks
old rank:     0
rank of x:    1
new rank:     1
greater than 0 - ok
```

This information disappears. The x moves from the top of the stack into place as the first character of the output expression. This move leaves the @ as the top (albeit only) item in the stack. Within the input expression, attention is still focused on the +. It is compared with the @ in the stack. The result is displayed:

```
compare:    +  @
----------------
precedence: 1  0
               >
             push
```

Thus, variables have precedence over + and + has precedence over @. The + sign is pushed onto the top of the stack.

```
|       |
|       |
|       |
|       |
|   +   |
|   @   |
```

The + sign passes from attention in the original expression, and the message to move forward reappears. You press the space bar again. The y in the original expression is now the focus of attention. You see it compared with the + that is now at the top of the stack.

```
compare:     y  +
----------------
precedence:  3  1
                >
              push
```

The y is pushed onto the stack.

When you press the space bar to move on, the @ at the end of the input expression comes under attention. Accordingly, the program pops the successive items from the stack until it reaches the @ at the bottom. First it checks that popping the y is valid:

```
popping - check ranks
old rank:    1
rank of y:   1
new rank:    2
greater than 0 - ok
```

The y is popped into the output expression, which becomes x y. The stack contents change to:

Then, after checking that the change in rank is valid, the program pops the + sign into the output expression, which becomes x y +. The rank is reduced to 1. The program finds that it has reached the @ at the bottom of the stack. Because the rank is 1 when the @ is reached, the conversion is correct.

The conversion of the infix expression a − b to its postfix form, ab −, is very similar to the conversion of a + b. The − symbol has the same precedence value as +. The conversion of a*b to ab* and the conversion of a/b to ab/ proceed in the same way, too. The precedence values of * and / are both 2.

The conversion of a + b*c to abc* + and the conversion of a*b + c to ab*c +, however, show the force of the algorithm when operators that have different precedence values are compared. The first of these conversions, that is, a + b*c to abc* +, can be summarized as follows:

Character under attention	a	+	+	b	*	*
Character currently at top of stack	@	a	@	+	b	+
Action	push	pop	push	push	pop	push
Contents of stack	@	a @	@	+ @	b + @	+ @
Output expression at this point			a	a	a	ab

Character under attention	c	@	@	@	@
Character currently at top of stack	*	c	*	+	@
Action	push	pop	pop	pop	
Contents of stack	* + @	c * + @	* + @	+ @	@
Output expression at this point	ab	ab	abc	abc*	abc* +

The conversion of a*b + c can be summarized correspondingly as follows.

a	*	*	b	+	+	+	c	@	@	@
@	a	@	*	b	*	@	+	c	+	@
push	pop	push	push	pop	pop	push	push	pop	pop	
@	a @	@	* @	b * @	* @	@	+ @	c + @	+ @	@
	a	a	a	ab	ab*	ab*	ab*	ab*c	ab*c +	

Both of these conversions produce postfix expressions in which the * is encountered before the +, regardless of their relative order in the infix expressions. This is because * has a higher precedence value than +, reflecting the fact that multiplication takes precedence over addition. A critical point occurs in the conversion of a + b*c when the * is under attention in the input and the + is at the top of the stack. The * is pushed onto the stack, because it has a higher precedence value than +. Inevitably, the * gets popped before the + later in the conversion. At the corresponding point in the conversion of a*b + c, however, the * is at the top of the stack and the + is under attention in the input. Consequently, the * is popped, and multiplication once more precedes addition in the evaluation of the result in the process that VALPOST demonstrates.

The conversion algorithm can be stated very simply.

1. Append an @ to the input expression and put @ into an otherwise empty stack. Set the rank to 0.

2. Perform steps 3 and 4 for each character of the input expression in turn, working from left to right until the @ is reached.

3. If the character under current attention in the input expression has higher precedence than the character at the top of the stack, then

push the current character from the input onto the stack and go on to the next character in the input.

4. Otherwise, add the rank of the current character in the input to the rank of the output formed so far. If the new rank is more than 0, pop the character at the top of the stack into the output and repeat step 3 by using the character that has just been "uncovered" as the new top character in the stack. If the new rank would be reduced to 0, however, quit with an error message.

5. When the @ that was appended to the input expression has been reached, perform step 6 until the @ at the bottom of the stack is reached.

6. Add the rank of the character at the top of the stack to the rank of the output formed so far. If the new rank is more than 0, pop the character at the top of the stack into the output.

7. When the @ at the bottom of the stack has been reached, accept the output expression as the correct result if its rank is 1. Otherwise, quit with an error message.

The treatment of expressions that contain too many operators is illustrated by $x + + y$. The processing ends when the output consists of the single term x, the first $+$ is in the stack, and the second $+$ is compared with the first $+$. The comparison of precedence values would pop the $+$ from the stack, but this would reduce the rank of the output to 0. The comment "less than 1, invalid expression" is displayed below the new rank, and the program prompts for another expression. The behavior when there are too many operands is shown by xy. The processing ends when the output consists of the terms x y and the stack contains only the @ symbol. The rank is found to be 2, and the comment "not equal to 1, invalid expression" is displayed beneath it.

Program Variables

T$(i)	The ith non-blank character in the input expression
ALPHABET$	The string of 52 characters consisting of all the lowercase letters a to z and capitals A to Z
D$	The dummy variable used to define the functions FN PCDNCE(D$) and FN RANK(D$) that give the precedence and rank, respectively, of an element in an expression
STC	The horizontal cursor position at which items in the stack are displayed
S$	The input expression stored as a string

I	The index that takes values 1, 2, . . . for successive non-blank characters in the input expression
IMAX	The maximum value of I, later increased by 1 to allow for the @ that the program puts at the end
J	The index of successive characters in the input expression
VO	The vertical position at which the input expression is redisplayed on the screen
HO	The horizontal position of the character under current attention in the input
VCP	The vertical position at which the postfix expression is displayed on the screen
HCP	The horizontal cursor position of the next character to be included in the postfix expression
RANK	The rank of the portion of the postfix expression that has been produced so far
STACK$(i)	The ith element currently in the stack
TOP	The number of entries currently in the stack, that is, the index of the topmost element
C$	The element to be pushed onto the stack
NIN$	The element under current attention in the infix expression
R$	The response to certain prompts
L	The line number in the FOR . . . NEXT loop that erases the message area

Program Notes

The application of the conversion algorithm begins by initializing RANK and TOP to 0 and planting an @ at the bottom of the empty stack. The main action is embodied in a FOR . . . NEXT loop that progresses through successive characters of the unpacked input expression until the terminal @ is reached.

```
1210 WHILE NIN$<>"@"
 . . .
1320 WEND
```

Within this loop, the program jumps to the subroutine at line 5500 to compare the current element of the infix expression with the topmost element of the stack. The critical statement is:

```
5580 IF FN PCDNCE(NIN$)>FN PCDNCE(STACK$(TOP)) THEN . . .
```

```
100 REM * POSTCON converts infix expressions to postfix *
101 :
110   ON ERROR GOTO 7000                                              'set error trap
120   KEY OFF: SCREEN 1: COLOR 1,0                                    'screen set up
130   DEF FN PCDNCE(D$)=-(INSTR("+-",D$)>0)-2*(INSTR("*/",D$)>0)
                          -3*(INSTR(ALPHABET$,D$)>0)                  'precedence
140   DEF FN RANK(D$)=(INSTR("+-*/",D$)>0)-(INSTR(ALPHABET$,D$)>0)    'rank
150   ALPHABET$="abcdefghijklmnopqrstuvwxyzABCDEFGHIJKLMNOPQRSTUVWXYZ"
160   DIM T$(40)                                                      'elements
170   STC=25                                                          'stack position
180   U$=SPACE$(40)                                                   'formatting string
199 :
200 REM * prompt for an expression *
210   CLS
220   LOCATE 4,1
230   PRINT "   Type a simple algebraic expression"                  'prompt for an
240   PRINT U$ "    that contains one-letter variables,"             'expression
250   PRINT U$ "        and +, -, * and / symbols,"
260   PRINT U$ "               for example"
270   PRINT U$ U$ "       a + b * c - d * p / q"
280   PRINT U$ U$ "   and then press the Enter key"
290   PRINT U$ " type here: ";
300   LINE INPUT S$
399 :
400 REM * separate the non blank characters *
410   IMAX=0
420   FOR I=1 TO LEN(S$)                                             'loop through expression
430     C$=MID$(S$,I,1)                                              'ith character
440     IF C$ = " " THEN 470
450       IMAX=IMAX+1
460       T$(IMAX)=C$                                                'transfer to array
470   NEXT I
480   IF IMAX < 20 THEN 610
490   PRINT U$ " Please use less than 20 letters and"               'too long
500   PRINT U$ " symbols. Press any key to try again.";
510   Z$=INPUT$(1)
520   GOTO 200                                                       'try again
599 :
600 REM * redisplay the original expression *
610   CLS                                                            'redisplay
620   PRINT U$ "Original expression: ";
630   IF IMAX>9 THEN PRINT U$ TAB(20-IMAX)                           'align or center
640   VO=CSRLIN                                                      'cursor positions
650   HO=POS(0)                                                      'for original and
660   VCP=2*VO                                                       'new expressions
670   HCP=HO
680   LOCATE VO,HO                                                   'redisplay
690   FOR I=1 TO IMAX
700     PRINT T$(I)+" ";                                            'ith character
710   NEXT I
720   PRINT U$ "Postfix expression: ";                              'caption for postfix
730   LOCATE 24,7
740   PRINT "Press space bar to go on";                             'prompt
750   Z$=INPUT$(1)
799 :
800 REM * draw the stack *
810   LINE (8*STC-20,72)-STEP (0,100)                                'left side of stack
820   LINE -STEP(32,0)                                               'base
830   LINE -STEP(0,-100)                                             'right side
999 :
```

```
1000 REM * convert the expression to postfix *
1010   RANK=0                                                    'initialize rank
1020   I=1                                             'pointers to infix expression
1030   TOP=0                                                      'and to stack
1040   C$="@"
1050   LOCATE 21,STC                                             'put @ in stack
1060   PRINT C$
1070   IMAX=IMAX+1                                      'put @ at end of expression
1080   T$(IMAX)="@"
1090   LOCATE VO,HO+2*IMAX-2                                        'display it
1100    PRINT T$(IMAX)
1110   GOSUB 6500                                              'wait for key press
1120   GOSUB 3000                                    'push Ith element onto stack
1130   GOSUB 5000                             'focus on next character of input
1199 :
1200 REM * loop through remaining elements *
1210   WHILE NIN$<>"@"                                 'loop through infix expression
1220   GOSUB 5500                                   'compare infix and stack items
1230    WHILE
           FN PCDNCE(NIN$)<=FN PCDNCE(STACK$(TOP))   'loop while stack dominates
1240      GOSUB 4000                                                  'pop
1250      GOSUB 5500                                 'compare input and stack items
1260    WEND
1270    C$=NIN$                                              'push onto stack
1280    GOSUB 3000
1290    GOSUB 6000                                         'clear the message area
1300    I=I+1
1310    GOSUB 5000                                     'focus on next input element
1320   WEND
1399 :
1400 REM * end of infix expression reached - empty the stack *
1410   WHILE STACK$(TOP)<>"@"                          'loop until marker is reached
1420    GOSUB 6500                                           'wait for key press
1430    GOSUB 4000                                                      'pop
1440   WEND
1442   IF RANK=1 THEN 1450                                      'jump if valid
1444   LOCATE 12,1: PRINT "old rank:   " RANK
1446   LOCATE 14,1: PRINT "not equal to 1"
1448   GOTO 6800                                             'jump to diagnostic
1450    C$="@"
1460    GOSUB 3110                                          'highlight the @
1470    GOSUB 6500                                          'wait for key press
1480    LOCATE VO,HO+2*IMAX-2
1490     PRINT SPC(1)                          'erase the @ from the infix expression
1500    REM
1510    REM
1520    POKE &H4E,3                                            'default color
1999 :
2000 REM * prompt for further action *
2010   LOCATE 24,1                                      'prompt for further action
2020    PRINT "(A)nother or (Q)uit" SPC(15)
2030   R$=INPUT$(1)
2040    IF R$="q" OR R$="Q" THEN 8000                                'branch
2050    GOTO 200
2999 :
3000 REM * push *
3010   IF TOP=0 THEN 3050
3020    POKE &H4E,3                                'de-highlight the top element
3030    LOCATE 22-TOP,STC
3040    PRINT STACK$(TOP)
```

```
3050    TOP=TOP+1                                          'increase the stack pointer
3060    STACK$(TOP)=C$                                        'store item in stack
3070    POKE &H4E,2                                               'highlight it
3080     LOCATE 22-TOP,STC
3090      PRINT C$
3100     IF TOP=1 THEN 3140
3110      POKE &H4E,1                          'de-highlight element in infix expression
3120      LOCATE VO,HO+2*I-2
3130      PRINT C$
3140    RETURN
3999  :
4000 REM * pop *
4010    C$=STACK$(TOP)                            'get element from top of stack
4020    POKE &H4E,3
4030     LOCATE 10,1                              'display rank check message
4040      PRINT "popping - check ranks"
4050      PRINT: PRINT "old rank:  "  RANK
4060      PRINT "rank of " C$ ": " FN RANK(C$)
4070    RANK=RANK+FN RANK(C$)                                 'compute new rank
4080    PRINT "new rank:  " RANK                               'display it
4090     IF RANK<1 THEN PRINT: PRINT "less than 1": GOTO 6800    'jump on error
4100      PRINT: PRINT "greater than 0 - ok"                    'rank is ok
4110      GOSUB 6500                                      'wait for key press
4120      GOSUB 6000                                      'clear message area
4130    LOCATE 22-TOP,STC                              'erase top item from stack
4140     PRINT SPC(1)
4150    TOP=TOP-1                                      'decrease stack pointer
4160    POKE &H4E,2                            'highlight the new top element
4170     LOCATE 22-TOP,STC
4180      PRINT STACK$(TOP)
4190    POKE &H4E,3                    'display popped element in postfix expression
4200     LOCATE VCP,HCP
4210      PRINT C$+" ";
4220     HCP=HCP+2
4230    RETURN
4999  :
5000 REM * get next character of input expression *
5010    GOSUB 6500                                          'wait for key press
5020    POKE &H4E,3
5030     LOCATE 10,1                                          'display comment
5035      PRINT "move forward"
5040    GOSUB 6500                                          'wait for key press
5050     LOCATE 10,1                                          'erase the comment
5055 LOCATE 10,1: PRINT SPC(15)
5060    NIN$=T$(I)                                          'get infix item
5070    POKE &H4E,2
5080     LOCATE VO,HO+2*I-2                            'highlight it in infix
5090      PRINT NIN$
5100    POKE &H4E,3
5110       RETURN
5499  :
5500 REM * compare characters in infix expression and stack *
5510    GOSUB 6500                                      'wait for key press
5520    LOCATE 10,1
5530     PRINT "compare:     " NIN$ SPC(2) STACK$(TOP)            'display message
5540     PRINT "----------------"
5550     GOSUB 6500
5560    PRINT USING "precedence:## ##"; FN PCDNCE(NIN$), FN PCDNCE(STACK$(TOP))
5570    GOSUB 6500
5580    IF FN PCDNCE(NIN$)> FN PCDNCE(STACK$(TOP)) THEN 5620 'compare precedences
```

```
5590    PRINT TAB(14) "<="
5600     PRINT TAB(13) "pop"                              'message to pop
5610     GOTO 5640
5620     PRINT TAB(14) ">"
5630     PRINT TAB(13) "push"                             'message to push
5640     GOSUB 6500
5650     GOSUB 6000
5660    RETURN
5999 :
6000 REM * clear message area *
6010    FOR L=10 TO 20                                    'clear the message area
6020     LOCATE L,1
6030      PRINT SPC(22)
6040    NEXT L
6050    RETURN
6499 :
6500 REM * wait for key press *
6510    IF INKEY$<>"" THEN 6510                           'clear keyboard buffer
6520     Z$=INPUT$(1)                                     'wait for key press
6530     IF R$=CHR$(27) THEN 8000                         'quit if Esc
6540    RETURN
6799 :
6800 REM * invalid input *
6810    LOCATE 18,1
6820     PRINT "invalid expression"                       'diagnostic
6830     GOTO 2000
6999 :
7000 REM * error trap *
7010    LOCATE 24,1                                       'system error
7020     PRINT "Error" ERR "in line" ERL "- press space bar";  'diagnostic
7030     Z$=INPUT$(1)
7040    RESUME 8000
7999 :
8000 REM * quit *
8010    POKE &H4E,3                                       'default foreground
8020     SCREEN 0: WIDTH 80                               'default screen
8030     LOCATE 22,1                                      'for Ok message
8040     ON ERROR GOTO 0                                  'reset error trap address
8050     END                                              'quit
```

This simply controls the displayed message that popping or pushing will occur. On return from the subroutine, the actual popping is performed or bypassed by an inner WHILE . . . WEND loop. This cycles while the precedence of the character under current attention in the infix expression is less than or equal to the precedence of the character at the top of the stack. Whenever popping starts, it continues for as long as this condition persists for the elements that surface as the topmost.

```
1230    WHILE FN PCDNCE (NIN$) <=FN PCDNCE (STACK$ (TOP) )
  . . .
1260    WEND
```

Within this inner loop, the subroutine at line 4000 is used to pop the top element from the stack into the postfix expression on the screen.

Following this loop, the subroutine at line 3000 pushes the item under current attention in the input expression onto the stack. Then the subroutine at line 5000 moves attention forward to the next character of the input.

When the @ has been reached at the end of the input, the items still in the stack are popped by another WHILE . . . WEND loop.

```
1410   WHILE STACK$(TOP)<>"@"
1430     GOSUB 4000
1440   WEND
```

The validation is performed by statements that test RANK in the popping subroutine and after the action that empties the stack.

Suggested Activities

1. Adapt POSTCON to convert (*a*) logical expressions and (*b*) set expressions into postfix form. (See Suggested Activities 1 and 2 in Section 4.3.)

2. Write a program to convert an arithmetical expression into prefix form. (See Suggested Activity 3 in Section 4.3.) Use a left-to-right scan, and use separate stacks for the variables and the operators.

3. An expression also can be converted to prefix form by reversing it, converting the result into postfix form, and reversing the result. Write a program to form prefix expressions this way.

4.5   Taking Account of Parentheses

The program PARENCON extends the action of POSTCON by allowing parentheses in the infix expression that is converted to postfix form. Thus a*(b + c) is converted to abc + *. The program also allows exponentiation, that is, raising to a power, denoted by ^. Thus a^b stands for a to the power b. The general pattern of behavior is very similar to that of POSTCON.

PARENCON begins by prompting:

```
Type a simple algebraic expression
that contains one-letter variables,
    +, -, *, /, and ^ symbols,
  and parentheses, for example,
    (a + b) * (c - d * p) / q
  and then press the Enter key
type here:
```

You respond, for example, by entering the expression x*(y + z). The program then works through this in an overall manner that is very similar to the behavior of POSTCON described in Section 4.3. Each character of the input expression is considered in turn and compared with the character at the top of the stack. Depending on the outcome of this comparison, the item is pushed onto the stack or one or more characters are popped from the top of the stack. To allow for parentheses, the algorithm associates three numbers with each character. These are the input precedence, the stack precedence, and the rank.

The basic cycle consists of comparing (1) the input precedence of the element under current attention in the input expressions with (2) the stack precedence of the top character currently in the stack. The input precedence and the stack precedence were equal for the same element in POSTCON. The values may be different, however, for the same element in PARENCON, as the following table shows.

Element	Variable	+ −	* /	^	(	)	@
Input precedence	7	1	3	6	9	0	−1
Stack precedence	8	2	4	5	0	0	−1
Rank	1	−1	−1	−1			

These precedence values lead to postfix expressions that correspond to left-to-right evaluation of infix expressions such as $a+b+c$ and $a*b*c$ but right-to-left evaluation of a^b^c. In other words, $a+b+c$ is treated as $(a+b)+c$ but a^b^c is treated as $a^{(b^c)}$. For addition and multiplication, $(a+b)+c$ is equal to $a+(b+c)$ and $(a*b)*c$ is equal to $a*(b*c)$. The value of $(a^b)^c$ need not be the same as $a^{(b^c)}$, however, even when a, b, and c are equal: Thus $(3^3)^3 = 27^3 = 19,683$, but $3^{(3^3)} = 3^{27} = 27^9 = 19,683^3$.

The program puts an extra pair of parentheses around the input expression at the start of the conversion. When a right parenthesis is under attention in the input, and a left parenthesis is at the top of the stack, both are discarded. The conversion of $x*(y+z)$ proceeds as follows. The program redisplays it encased in extra parentheses as $(x*(y+z))$. The successive cycles (1) push the (onto the empty stack, (2) push the x onto the stack, (3) pop the x into the output, (4) push the * onto the stack; (5) push the second (onto the stack, (6) push the y onto the stack, (7) pop the y into the output, (8) push the + onto the stack, (9) push the z onto the stack, (10) pop the z into the output. Then, in step 11, the first) is reached in the input and the + is popped. This makes the output xyz+ and exposes the second (in the stack. In step

12, the) under attention in the input and the (at the top of the stack are discarded. Then, in step 13, attention moves to the second) in the input. The * is at the top of the stack. It is popped, exposing the (at the bottom of the stack. In step 14, the) under attention at the end of the input expression and the (at the bottom of the stack are discarded. The stack is empty and the final character of the input has been processed. Therefore, the conversion is complete.

input	(x* (y+z))	(x* (y+z))	(x* (y+z))	(x* (y+z))
partial output			x	x
stack	((1)	x ((2)	((3)	* ((4)

input	(x* (y+z))	(x* (y+z))	(x* (y+z))	(x* (y+z))
partial output	x	x	x y	x y
stack	(* ((5)	y (* ((6)	(* ((7)	+ (* ((8)

input	(x* (y+z))	(x* (y+z))	(x* (y+z))	(x* (y+z))
partial output	x y	x y z	x y z +	x y z +
stack	z + (* ((9)	+ (* ((10)	(* ((11)	* ((12)

input	(x* (y+z))	(x* (y+z))
partial output	x y z + *	x y z + *
stack	(	
	(13)	(14)

Because the program puts an extra pair of parentheses around the input, the process always ends in the same way as this example when the input is valid. Thus, if the expression consists of the single character x, it is redisplayed as (x). The (is pushed onto the stack. The x is put into the output. The) that follows it in the modified input matches the (at the bottom of the stack. Both are discarded, leaving the stack empty and the input fully processed.

If the input contains too many right parentheses, the stack empties before the end of the expression has been reached. If the input contains too few right parentheses, the end of the expression is reached before the stack is empty. Other errors are detected by a pop action that makes the rank negative. The reader should trace the behavior of several invalid expressions as well as several expressions which are correct. It is essential for an algorithm such as this to recognize invalid expressions as well as processing valid ones correctly.

Program Variables

The program PARENCON uses the same variables as POSTCON and also

G   The switch that is 0 except when a) in the output expression and a (at the top of a stack have just been discarded, at which point G is switched to −1 until the next element has been popped from the stack.

Program Notes

The general operation of PARENCON follows the pattern set by POSTCON. The character-by-character progression through the infix expression is handled by a WHILE . . . WEND loop that ends when I reaches IMAX. An inner WHILE . . . WEND loop controls the overall

```
1ØØ REM * PARENCON converts infix expressions with parentheses to postfix *
1Ø1 :
11Ø   ON ERROR GOTO 7ØØØ                                      'set error trap
12Ø    KEY OFF: SCREEN 1: COLOR 1,Ø                           'screen set up
13Ø   DEF FN INPCD(D$)=VAL(MID$(INPCD$,INSTR(CHARSET$,D$)+1,1))
                   + (INSTR(CHARSET$,D$)=Ø)                   'infix precedence
14Ø   DEF FN STPCD(D$)=VAL(MID$(STPCD$,INSTR(CHARSET$,D$)+1,1))
                   + (INSTR(CHARSET$,D$)=Ø)                   'stack precedence
15Ø   DEF FN RANK(D$)=SGN(INSTR(CHARSET$,D$)-5.5)                      'rank
16Ø   ALPHABET$="abcdefghijklmnopqrstuvwxyzABCDEFGHIJKLMNOPQRSTUVWXYZ"
17Ø    CHARSET$="+-*/^"+ALPHABET$+"()"
18Ø    INPCD$="Ø11336"+STRING$(52,"7")+"9Ø"       'infix precedence values
19Ø    STPCD$="Ø22445"+STRING$(52,"8")+"Ø"                 'stack values
2ØØ   DIM T$(4Ø)                                  'elements of infix expression
21Ø   STC=25                                               'stack position
22Ø   U$=SPACE$(4Ø)                                         'formatting string
299 :
3ØØ REM * prompt for an expression *
31Ø   CLS
32Ø   LOCATE 4,1
33Ø   PRINT "   Type a simple algebraic expression"         'prompt for an
34Ø   PRINT U$ "   that contains one-letter variables,"      'expression
35Ø   PRINT U$ "        +, -, *, / and ^ symbols,"
36Ø   PRINT U$ "      and parentheses, for example"
37Ø   PRINT U$ U$ "       (a + b) * (c - d * p )/ q"
38Ø   PRINT U$ U$ "     and then press the Enter key"
39Ø   PRINT U$ " type here:   ";
4ØØ    LINE INPUT S$
499 :
5ØØ REM * separate the non blank characters *
51Ø   IMAX=1                                              'starting parenthesis
52Ø    T$(1)="("
53Ø   FOR I=1 TO LEN(S$)                                  'loop through expression
54Ø    C$=MID$(S$,I,1)                                    'ith character
55Ø    IF C$ = " " THEN 59Ø
56Ø     IF INSTR(CHARSET$,C$)=Ø THEN GOTO 58ØØ            'invalid character
57Ø      IMAX=IMAX+1
58Ø      T$(IMAX)=C$                                       'transfer to array
59Ø   NEXT I
6ØØ   IMAX=IMAX+1                                          'closing parenthesis
61Ø    T$(IMAX)=")"
62Ø   IF IMAX < 2Ø THEN 71Ø
63Ø   PRINT U$ " Please use less than 19 letters and"               'too long
64Ø   PRINT U$ "  symbols. Press any key to try again.";
65Ø    Z$=INPUT$(1)
66Ø    GOTO 3ØØ
699 :
7ØØ REM * redisplay the original expression *
71Ø   CLS
72Ø   PRINT U$ "Original expression: ";
73Ø   IF IMAX>9 THEN PRINT U$ TAB(2Ø-IMAX)                   'align or center
74Ø   VO=CSRLIN                                             'cursor positions
75Ø   HO=POS(Ø)                                             'for original and
76Ø   VCP=2*VO                                              'new expressions
77Ø   HCP=HO
78Ø   LOCATE VO,HO                                          'redisplay
79Ø   FOR I=1 TO IMAX
8ØØ    PRINT T$(I)+" ";                                     'ith character
81Ø   NEXT I
82Ø   PRINT U$ "Postfix expression:   ";                    'caption for postfix
```

```
83Ø    LOCATE 24,7
84Ø    PRINT "Press space bar to go on";                          'prompt
85Ø    Z$=INPUT$(1)
899 :
9ØØ REM * draw the stack *
91Ø    LINE (8*STC-2Ø,72)-STEP (Ø,1ØØ)                   'left side of stack
92Ø    LINE -STEP(32,Ø)                                               'base
93Ø    LINE -STEP(Ø,-1ØØ)                                       'right side
999 :
1ØØØ REM * begin the conversion *
1Ø1Ø   RANK=Ø                                             'initialize rank
1Ø2Ø   I=1                                       'pointers to infix expression
1Ø3Ø   TOP=Ø                                                   'and stack
1Ø4Ø   C$=T$(I)
1Ø5Ø   GOSUB 55ØØ                                        'wait for key press
1Ø6Ø   GOSUB 2ØØØ                                'push opening ( onto stack
1Ø7Ø   I=I+1
1Ø8Ø   GOSUB 4ØØØ                        'get next element of infix expression
1199 :
12ØØ REM * loop through remaining items *
121Ø   WHILE I<=IMAX                              'loop through successive items
1215   IF TOP<1 THEN 58ØØ
122Ø   GOSUB 45ØØ                              'compare infix and stack items
123Ø    WHILE FN INPCD(NIN$)<=FN STPCD(STACK$(TOP)) 'loop while stack dominates
124Ø     GOSUB 3ØØØ                                                   'pop
125Ø     IF NOT G THEN 13ØØ                   'jump unless paired ) & ( discarded
126Ø      G=Ø                                            'reset the switch
127Ø      C$=NIN$
128Ø      GOSUB 21ØØ                          'lowlight item in infix expression
129Ø      GOTO 135Ø                              'jump out of popping loop
13ØØ      GOSUB 45ØØ                            'compare infix and stack items
131Ø    WEND
132Ø    C$=NIN$
133Ø    GOSUB 2ØØØ                                     'push item onto stack
134Ø    GOSUB 5ØØØ                                     'clear the message area
135Ø    I=I+1
136Ø     GOSUB 4ØØØ                                    'get next infix item
137Ø   WEND
1499 :
15ØØ REM * expression has been processed *
151Ø   IF TOP>Ø OR RANK<>1 THEN 58ØØ                      'check for validity
153Ø   LOCATE 24,1
154Ø    PRINT "(A)nother or (Q)uit" TAB(39)            'prompt for further action
155Ø    R$=INPUT$(1)
156Ø    IF R$="q" OR R$="Q" THEN 8ØØØ                             'branch
157Ø    GOTO 3ØØ
1999 :
2ØØØ REM * push *
2Ø1Ø   IF TOP=Ø THEN 2Ø5Ø
2Ø2Ø     POKE &H4E,3                             'de-highlight the top element
2Ø3Ø     LOCATE 22-TOP,STC
2Ø4Ø     PRINT STACK$(TOP)
2Ø5Ø   TOP=TOP+1                                     'increase the stack pointer
2Ø6Ø   STACK$(TOP)=C$                                    'store item in stack
2Ø7Ø   POKE &H4E,2                                             'highlight it
2Ø8Ø   LOCATE 22-TOP,STC
2Ø9Ø   PRINT C$
21ØØ   POKE &H4E,1                        'de-highlight item in infix expression
211Ø   LOCATE VO,HO+2*I-2
212Ø    PRINT C$
```

```
2130   RETURN
2999 :
3000 REM * pop *
3005   IF TOP<1 THEN 5800
3010   C$=STACK$(TOP)                                   'get element from top of stack
3020   POKE &H4E,3
3030   LOCATE 10,1
3040    IF FN INPCD(NIN$)<FN STPCD(C$) THEN 3080        'jump unless paired ) & (
3050     PRINT "discard the ( and )"                         'discard if found
3060     G=-1                                                   'set switch
3070     GOTO 3150
3080   PRINT "popping - check ranks"                    'display rank checking message
3090     PRINT: PRINT "old rank:   " RANK
3100     PRINT "rank of " C$ ": " FN RANK(C$)
3110   RANK=RANK+FN RANK(C$)                                 'compute new rank
3120     PRINT "new rank:   " RANK                           'display it
3130     IF RANK<1 THEN GOTO 5800                            'jump on error
3140     PRINT: PRINT "greater than 0 - ok"                   'rank is ok
3150     GOSUB 5500                                       'wait for key press
3160     GOSUB 5000                                       'clear message area
3170   LOCATE 22-TOP,STC                                 'erase top item from stack
3180     PRINT SPC(1)
3190   TOP=TOP-1                                         'decrease stack pointer
3200   POKE &H4E,2                                       'highlight new top element
3210   LOCATE 22-TOP,STC
3220     PRINT STACK$(TOP)
3230   IF G THEN 3280                                    'jump if ( was popped
3240   POKE &H4E,3                          'display popped item in postfix string
3250   LOCATE VCP,HCP
3260     PRINT C$+" ";
3270     HCP=HCP+2
3280   RETURN
3999 :
4000 REM * get next character of input expression *
4010   GOSUB 5500                                            'wait for key press
4020   POKE &H4E,3
4030   LOCATE 10,1
4040    PRINT "move forward";                               'display comment
4050   GOSUB 5500
4060   IF I<=IMAX THEN 4090
4070    PRINT " - end has been reached"                       'end reached
4080    RETURN
4090   LOCATE 10,1                                           'erase comment
4100    PRINT SPC(15)
4110   NIN$=T$(I)                                           'get next infix item
4120   POKE &H4E,2                               'highlight it in infix expression
4130   LOCATE VO,HO+2*I-2
4140    PRINT NIN$
4150   POKE &H4E,3                                           'default color
4160   RETURN
4499 :
4500 REM * compare characters in infix expression and stack *
4510   GOSUB 5500                                            'wait for key press
4520   LOCATE 10,1
4530    PRINT "compare:    " NIN$ SPC(2) STACK$(TOP)         'display message
4540    PRINT "----------------"
4550   GOSUB 5500
4560   PRINT USING "precedence:## ##"; FN INPCD(NIN$), FN STPCD(STACK$(TOP))
4570   GOSUB 5500
4580   IF FN INPCD(NIN$)> FN STPCD(STACK$(TOP)) THEN 4620    'compare precedence
```

```
4590    PRINT TAB(14) "<="
4600    PRINT TAB(13) "pop"                           'message to pop
4610    GOTO 4640
4620  PRINT TAB(14) ">"
4630    PRINT TAB(13) "push"                          'message to push
4640    GOSUB 5500
4650    GOSUB 5000
4660  RETURN
4999 :
5000 REM * clear display area *
5010  FOR L=10 TO 20                                  'clear the message area
5020    LOCATE L,1
5030      PRINT SPC(22)
5040  NEXT L
5050  RETURN
5500 REM * wait for key press *
5510  IF INKEY$<>"" THEN 5510                         'clear keyboard buffer
5520    Z$=INPUT$(1)                                  'wait for key press
5530  IF R$=CHR$(27) THEN 8000                        'quit if Esc
5540  RETURN
5799 :
5800 REM * invalid input *
5810  LOCATE 16,1
5820    PRINT "invalid expression"                    'diagnostic
5830    GOTO 1530
6999 :
7000 REM * error trap *
7010  LOCATE 24,1                                     'system error
7020    PRINT "Error" ERR "in line" ERL "- press space bar";   'diagnostic
7030    Z$=INPUT$(1)
7040  RESUME 8000
7999 :
8000 REM * quit *
8010  POKE &H4E,3                                     'default foreground
8020  SCREEN 0: WIDTH 80                              'default screen
8030  LOCATE 22,1                                     'for Ok message
8040  ON ERROR GOTO 0                                 'reset error trap address
8050  END                                            'quit
```

popping process. A few simple subroutines move attention in the infix
string and perform the push and pop actions.

```
1210   WHILE I<=IMAX
. . .
1230     WHILE FN INPCD(NIN$) <= FN STPCD(STACK$(TOP))
. . .
1310     WEND
. . .
1370   WEND
```

The functions FN INPCD$(D$), FN STPCD$(D$), and FN RANK(D$)
return the input precedence, stack precedence, and rank of D$.

Suggested Activities

1. Adapt PARENCON (*a*) to convert logical expressions that may contain parentheses and (*b*) to convert set expressions that may contain parentheses into postfix form.

2. Write a program to convert an arithmetic expression that may contain parentheses into prefix form.

3. Write a program to convert a prefix expression into infix form.

4. Write a program to convert a postfix expression into infix form.

4.6 Servicing a Queue

A stack operates on a last in, first out basis, often abbreviated LIFO. This is the opposite of the first in, first out or FIFO principle that is used to service waiting lines in cafeterias and banks and innumerable other situations. Engineers and analysts who are concerned with the flow of traffic use computers to monitor and to simulate the behavior of the systems that they design. The word "traffic" is used in a very general sense—automobiles on streets and highways, aircraft around an airport, customers in a department store, mail that is being sorted, and so on. In these simulation studies, it is necessary to keep track of the changing state of a waiting line such as the identities of the airplanes waiting to land in their order of arrival in the airspace around an airport.

One way to keep track of that waiting line is to use an array large enough to store the identities of all the planes that can land on a single day. At the end of the day, the program is restarted by carrying over the data about planes still in the air. This is not very convenient, and it is also quite unnecessary. All that you need is an array large enough to hold the data concerning the greatest number of planes that may be in the air at the same time.

When the airport opens, you store the data about the first arrival as the first element of the array. Whenever another plane arrives, you store the data about it as the element one place on from the element that stores the data about the preceding arrival. Whenever a plane has landed, you delete the information about it. Sooner or later the new arrivals in the airspace reach the top of the array, but by then the planes that landed have freed space at the bottom of the array. (The "top" elements have the highest subscript values, and the "bottom" elements have the lowest values, regardless of whether the contents of the array are displayed with subscript values increasing or decreasing in the field of view.) Consequently, you can start putting new information at the bottom and carry on in this way indefinitely. The term "wrap-around" is used for this tactic of continuing cyclically from the top of an array to the bottom. It works provided you can keep track of where the unprocessed data starts and ends.

The program QUEUE demonstrates the "cyclic queue" or wrap-around principle that has just been described. The program relates to a slightly different "plot mechanism"—shoppers arriving in a store that has one "server," that is, one salesperson dealing with their purchases. The server deals with only one customer at a time and goes behind the scenes, intermittently, for random periods of time. The program begins by displaying

1. The twenty slots into which the initials of shoppers will be inserted
2. The prompt that tells you to press the space bar to make the program cycle through successive steps of the simulation.

```
          1 ---
          2 ---
           . . .
         19 ---
         20 ---

Press space bar to continue
```

You respond, and the program incorporates a random selection of three letters and the word "his" or "her" in a message at the left of the screen. If, for example, the random initials are RMZ and the selected word is "her," the display changes to:

```
RMZ wants to      1 ---
join the queue.   2 ---
                  3 ---
Put her name      4 ---
in slot 1.        5 ---
                   . . .
                 20 ---
```

You press the space bar again. The left side of the screen clears, and the top line of the queue changes to

```
          1 RMZ←head of queue
```

You press the space bar again and the message containing random initials and "his" or "her" reappears, mentioning slot 2 as the destination. The message could read, for example, "JPD wants to join the queue. Put her name in slot 2."

Pressing the space bar again then changes the top of the display to

```
1 RMZ←head of queue
2 JPD←tail of queue
```

After a shopper has joined the queue, the program decides randomly whether the next action will be the arrival of another shopper or the reappearance of the server. When the queue is very small, this random decision is loaded in favor of a shopper joining the queue (passersby see that they will be served quickly and come to take advantage of the lull). Typically, this may go on say for four cycles, leading to a display such as

```
1 RMZ←head of queue
2 JPD
3 IGX
4 AVA←tail of queue
5 ---
. . .
```

At this point the program is quite likely to service a customer when you next press the space bar. The display then changes to

```
Server is ready.    1 RMZ←head of queue
                    2 JPD
Slot 1 holds        3 IGX
the name of the     4 AVA←tail of queue
next person to      5 ---
be served           6 ---
                    7 ---
Serve RMZ           8 ---
                    . . .
```

The message disappears from the left side of the screen and the top three lines of the display change to

```
1 ---
2 JPD←head of queue
3 IGX
```

You continue to press the space bar and the program randomly:

1. Puts names into slots further and further down the display, each time moving the words "tail of queue" to be alongside the most recent arrival

2. Removes names from slots 2, 3, . . . in turn, each time moving the words "head of queue" to be alongside the earliest outstanding arrival

This leads sooner or later to the situation in which several slots reaching to the twentieth are occupied and another person wishes to join the queue. The display at that point might be:

```
OVZ wants to        1 ---
join the queue.     2 ---
                    3 ---
Put her name        4 ---
in slot 1.          5 ---
                     . . .
                    11 SRN←head of queue
                     . . .
                    20 ANR←tail of queue
```

Pressing the space bar again causes "wraparound":

```
                     1 OVZ←tail of queue
                     2 ---
                      . . .
                    11 SRN←head of queue
                      . . .
                    20 ANR
```

The program thus starts to reuse the top slots, which were vacated by earlier cycles in which the server dealt with customers. By continuing to press the space bar, you make the program randomly:

1. Put more initials into the upper part of the display
2. Remove initials from the lower part

The random selection of actions is weighted to prevent the tail from "catching up" with the head. Sooner or later, the head is at the twentieth slot and the server is ready to service the next customer. At this point the appearance of the display might be:

```
Server is ready.    1 OVZ
                    2 DHA
Slot 20 holds       3 PGG
the name of the     4 VQW
next person to      5 HNH
be served           6 AGS←tail of queue
                    7 ---
Serve ANR           8 ---
                     . . .
                    19 ---
                    20 ANR←head of queue
```

Pressing the space bar clears the bottom slot and moves the head of the queue to the top slot.

```
 1 OVZ←head of queue
 . . .
 6 AGS←tail of queue
 7 ---
 . . .
20 ---
```

The program continues in this manner for as long as you like. The occupied area migrates downward, wraps around, and coalesces at the top, over and over again. You press the Escape key to make the program terminate.

Program Variables

NLIM	The maximum length of the queue
A$(n)	The initials currently in the nth slot of the array that represents the queue
NLO	The index of the slot that corresponds to the current head of the queue
NHI	The index of the slot that corresponds to the current tail of the queue
NHIP	The previous value of NHI
FULL	The flag that has the value 1 when the queue is full and 0 otherwise
STARTUP	The flag that has the value 2 when the queue is empty, 1 when one person is in it, and 0 when two or more people are in it
ESC$	The one-byte string that contains the escape code
N	The index of A$ in several loops
Z$	The one-character response to several prompts
L$	The user-defined function that generates a randomly selected capital letter
LATEST$	The random initials of the latest arrival in the queue
FN U$	The user-defined function that moves the text cursor back to the left edge of the screen and then down a line to provide double spacing

Program Notes

The program begins with some simple initialization actions. These include defining the function FN L$ to form the character that has, as its ASCII code, a randomly selected integer in the range 65 to 90 that correspond, respectively, to the letters A to Z.

```
140 DEF FN L$ = CHR$(65+INT(26*RND))
```

The simulation cycle is embodied in a WHILE . . . WEND loop that terminates when the computer tries to start a new cycle after you press the Escape key.

```
1010    WHILE Z$<>ESC$
1020      Z$=INPUT$(1)
. . .
3180    WEND
```

A major branch within this loop selects enqueueing or dequeueing as the action of the current cycle. The selection is random but weighted by the occupancy of the array, that is, by the number of unserved customers. This number is simply NHI − NLO when the queue is not wrapped around, that is, when NHI>NLO. It is NLIM + NHI − NLO when the queue is wrapped around, that is, NHI<NLO. Both cases are subsumed by the expression

```
1 + ((NLIM + NHI - NLO) mod NLIM)
```

If you need assurance that this is correct, consider the cases

1. NHI = NLO
2. NHI = NLO + 1
3. NLO = NLIM, NHI = 1
4. NHI = NLO − 1
5. NLO = 1, NHI = NLIM

The occupancy is 1 in case 1, 2 in cases 2 and 3, and NLIM in cases 4 and 5.

The enqueue or dequeue action is determined by the branching statement:

```
1030 IF RND < ((NHI+NLIM-NLO) MOD NLIM)/NLIM THEN 3000
```

In this statement the value of the expression following the < symbol ranges from close to 0 to close to 1, as the queue changes from emptiness to saturation. The RND function produces a random number

```
100 REM * QUEUE demonstrates a circular queue *
101 :
110    ON ERROR GOTO 4000                                          'set error trap
120    KEY OFF: SCREEN 1                                           'screen set up
130      COLOR 1,0: CLS
140    DEF FN L$ = CHR$(65+INT(26*RND))                            'random letter
150    DEF FN U$=STRING$(POS(0)-1,29)+CHR$(31)                     'formatting function
160    NLIM=20                                                     'queue capacity
170      DIM A$(NLIM)
180    NLO=1                                                       'current head
190    NHI=1                                                       'current tail
200    NHIP=1                                                      'previous tail
210    STARTUP=2                                                   'queue is empty
220    FULL=0                                                      'queue is not full
230    ESC$=CHR$(27)                                               'Esc code
240    VIEW (0,0)-(150,180)                                        'message area
250      FOR N=1 TO 20                                            'display the schematic
260      LOCATE N,20
270      PRINT USING "## ---"; N
280      NEXT N
290    LOCATE 25,1                                                 'prompt
300      PRINT "Press space bar to continue.";
999 :
1000 REM * main cycle *
1010    WHILE Z$ <> ESC$                                          'loop until Esc pressed
1020      Z$=INPUT$(1)                                            'store a key press
1030      IF RND < ((NHI+NLIM-NLO) MOD NLIM)/(NLIM) THEN 3000     'randomly dequeue
1999 :
2000 REM * enqueue *
2010    IF FULL THEN PRINT "queue is full": GOTO 3180             'jump if full
2020      LATEST$=FNL$+FNL$+FNL$                                  'random initials
2030    CLS                                                       'clear message area
2040    LOCATE 6,1
2050      PRINT LATEST$ " wants to"                               'prompt to enqueue
2060      PRINT FNU$ "join the queue."
2070      PRINT FNU$ FNU$ "Put "
                 MID$("hisher",1+3*INT(2*RND),3) " name"          'random gender
2080      PRINT FN U$ "in slot" NHI CHR$(29) "."
2090    Z$=INPUT$(1)
2100    A$(NHI)=LATEST$                                           'put initials in slot
2110      LOCATE NHIP,26
2120      IF STARTUP=0 THEN PRINT SPC(14)                         'erase previous tail marker
2130      IF STARTUP<>1 THEN 2160
2140      PRINT CHR$(27) "head of queue"                          'queue has 2nd member - label head
2150        STARTUP=0
2160      LOCATE NHI,23
2170      PRINT A$(NHI) CHR$(27) "tail of queue"                  'label the tail
2180      IF STARTUP<>2 THEN 2220
2190        LOCATE NLO,27                                         'queue has 1st member
2200        PRINT "head"                                          'label the head
2210        STARTUP=1                                             'change the switch
2220      NHIP=NHI                                                'reset prior tail
2230      NHI=1+NHI MOD NLIM                                      'reset pointer to next free slot
2240      IF NHI=NLO THEN FULL=1                                  'test whether full
2250      GOTO 3180
2999 :
3000 REM * dequeue *
3010    CLS                                                       'clear message area
3020    LOCATE 5,1
3030      PRINT "Server is ready." FNU$                           'prompt to dequeue
```

```
3040    IF STARTUP=2 THEN PRINT "Queue is empty.": GOTO 300        'empty
3050    PRINT "Slot" NLO "holds"
3060      PRINT FNU$ "the name of the"              'describe dequeuing action
3070      PRINT FNU$ "next person to"
3080      PRINT FNU$ "be served."
3090      PRINT FNU$ "Serve " A$(NLO) "."
3100    Z$=INPUT$(1)                                    'wait for key press
3110    LOCATE NLO,23
3120      PRINT "---" SPC(14)                         'show slot is empty
3130    NLO=1+NLO MOD NLIM                          'reset pointer to head
3140    LOCATE NLO,26
3150      PRINT CHR$(27) "head of queue"              'label the head
3160    IF NLO=NHI THEN STARTUP=2                      'queue is empty
3170    IF FULL THEN FULL=0                             'queue is full
3180    WEND
3190    GOTO 5000
3999 :
4000 REM * error trap *
4010    PRINT "Error" ERR "in line" ERL "- press space bar";      'diagnostic
4020    Z$=INPUT$(1)                                'wait for key press
4030    RESUME 5000
4999 :
5000 REM * quit *
5010    REM
5020    ON ERROR GOTO 0                          'reset error trap address
5030    END                                                     'quit
```

with uniform probability of being between 0 and 1. The likelihood of the test being satisfied and the program branching to the dequeueing action thus is very low when the queue is empty and very high when it is full.

In the enqueueing action, the random initials are constructed by concatenating three references to the random letter function FN L$.

Suggested Activities

1. Adapt the program QUEUE for interactive use. Make the basic operations (*a*) adding a name to the end of the queue, (*b*) removing a name from the end of the queue, and (*c*) finding the number of names currently in the queue.

2. Extend Activity 1 to allow the removal of the name closest to the head of the queue that begins with a letter that the user specifies. This can create free slots between slots that are occupied. Make the program incorporate these free slots in the free space during wraparound.

3. Rework Activity 2 to use a linked list representation of the queue.

4. If you can add and remove objects from both the beginning and the end of a queue, it is called a double queue (deque). Adapt the pro-

gram QUEUE to demonstrate the operation of a deque that is stored in an array by using wraparound.

5. Expand Activity 1 to maintain several queues in the same array by using linked list representation.

6. Adapt the program QUEUE to simulate a situation in which (1) the shoppers arrive at random time intervals that are uniformly distributed in the range 5 to 15 seconds and (2) the server takes 2 to 18 seconds to deal with a customer, with an average serving time of 10 seconds. Use the TIMER function.

7. Extend Activity 6 to allow other statistical distributions of arrival interval and serving time.

4.7 Some Further Directions

Throughout this chapter, stacks and queues have been stored serially in arrays. Other methods, such as linked lists, are useful at times. Stacks and queues can be regarded as abstract data types, in the sense discussed in Section 3.5, independent of the method of storage. Thus, the following set of commands will support any algorithm that uses stacks regardless of how they are stored: (1) create an empty stack, (2) push data onto the stack, (3) find whether the stack is empty, (4) pop data from the stack and, at times, (5) find the current depth of the stack, and (6) peep at the data at a specified distance below the surface. Computers have, in fact, been built with special units to store data in stacks that can be controlled by the commands which were just described.

The use of stacks to deal with digressions is particularly important in applications that involve "recursion." The general idea of recursion is contained in the jingle "Big fleas have little fleas upon their backs to bite them, and little fleas have lesser fleas and so on ad infinitum." Devotees of Dr. Seuss encounter recursion when the "cat in the hat" doffs his hat, revealing smaller cats in hats who doff their hats revealing still smaller cats, and so on.

The idea of recursion is very important when you are dealing with certain kinds of patterns. Thus, the structure of algebraic expressions that contain + and × symbols and parentheses can be described recursively as follows. A sum consists of two or more terms joined by + symbols. A product consists of two or more factors joined by × symbols. A parenthesized expression consists of an algebraic expression with parentheses around it. Each term in a sum may be a number or a variable or a product or a parenthesized expression. Each factor in a product may be a number or a variable or a parenthesized expression. An algebraic expression may be a sum or a product or a parenthesized expression.

This definition is cyclic in the sense that it defines an algebraic expression in terms of simpler algebraic expressions. Thus the expression $(a + (b + c) \times (d + e + f) + g \times h)$ fits the definition as follows. The letters a, ..., h are variables, in accordance with accepted conventions. Consequently, each can serve as a term in a sum or as a factor in a product. Consequently, $b + c$ and $d + e + f$ are valid sums and $g \times h$ is a valid product. Because $b + c$ is a valid sum, it is a valid algebraic expression. Therefore, putting parentheses around it makes $(b + c)$ a valid parenthesized expression. Likewise, because $d + e + f$ is a valid sum, it is a valid algebraic expression. Therefore, $(d + e + f)$ is a valid parenthesized expression. Because $(b + c)$ and $(d + e + f)$ are valid parenthesized expressions, they can serve as factors in a product. Consequently, $(b + c) \times (d + e + f)$ is a valid product. Now consider the expression $a + (b + c) \times (d + e + f) + g \times h$. Because a is a variable, it can serve as a term in a sum. Because $(b + c) \times (d + e + f)$ and $g \times h$ are valid products, they can serve as terms in a sum. Consequently, $a + (b + c) \times (d + e + f) + g \times h$ is a valid sum. Therefore, it is a valid algebraic expression. Therefore, $(a + (b + c) \times (d + e + f) + g \times h)$ is a valid parenthesized expression that can be used, if necessary, in still longer expressions. This particular example thus consists of a parenthesized expression which contains subsidiary parenthesized expressions.

We can define the structure of algebraic expressions recursively when prefix and postfix notation is used, too. Thus, in prefix notation, an algebraic expression is either (1) a number or (2) a variable or (3) a binary operator followed by two algebraic expressions. The corresponding definition for postfix notation simply uses "preceded" instead of "followed" in alternative 3. The evaluation of postfix expressions by the VALPOST program in Section 4.3 used a stack to allow for the recursive structure of the expressions. So did the conversion of unparenthesized and parenthesized expressions to postfix form by the POSTCON and PARENCON programs in Sections 4.4 and 4.5. We mentioned recursion, too, in connection with the quicksort program in Section 1.4. That algorithm sorts a list by moving an element into its final position and then sorting the sublists on each side in the same way. The algorithm terminates because, sooner or later, each sublist that still requires sorting contains a single element.

The idea of recursion has many uses, and it can be embodied in programs that have many applications. For example, you may want to produce an endless supply of algebraic expressions for students to evaluate or to simplify. The recursive description of the structure of such expressions that was given earlier in this section can be used to generate the examples mechanically. It can also be used to recognize whether a given expression satisfies the rules for correct formation. The recursive definition and implementation of algorithms is particularly important when you are dealing with trees and graphs of information, the subject of Chapter 5.

How to Make Connections

Preceding chapters have dealt largely with data that had a natural numeric or alphabetic order or that had a predetermined order of arrival in a situation that was changing dynamically. We turn our attention now to bodies of data that have multiple interconnections. Thus, commerce is connected by vendor-buyer relationships, bones are connected by muscles and tendons, molecules consist of atoms that are joined by forces of valency, and so on.

Modern computers have extended the domain of mechanical information processing from data that just had neighbors in a one-dimensional sequencing to data joined by more extensive relationships. Road maps, organization charts, flow diagrams, family trees, and many other kinds of diagram are used to show how objects are joined. All these diagrams represent "graphs" in the sense that will now be explained. A graph can be depicted by a collection of (1) points that represent different objects (such as towns or people or actions) and (2) lines that join particular pairs of points because of a relationship between them. Thus, an air travel map contains lines that correspond to direct flights between pairs of towns. A family tree contains lines between parents and children. A molecular diagram contains lines between atoms that are bonded together. A flow chart contains lines between actions that can occur in succession.

Each of the points that represents an object in a graph is called a *node*. Each line that joins a pair of nodes is called an *edge*. Other terms are used, too, such as "vertex" for node and "arc" for edge. If it is necessary to associate a direction with each edge—for example, in a network of one-way streets—the graph is said to be *directed*. Otherwise, it is *undirected*. Sometimes, all that is of interest is whether or not particular pairs of nodes are connected; sometimes, the distance between nodes, or some equivalent numerical property, is of concern. The numbers associated with the edges then are called *lengths* or *weights,* and the graph is said to be *weighted.*

When it is possible to get from every node in a graph to every other node by one and only one path, the graph is called a *tree.* Thus, a family tree that shows just the males or just the females is a tree in the mathematical sense. So is an organization chart in which each person has a single immediate supervisor.

Computers have been used, to date, to deal with trees and graphs of data of direct interest to end users in several kinds of application, such as traffic flow, econometric optimization, genealogy, and chemical structure. Trees and graphs also are used extensively in specialized algorithms for processes, such as sorting and searching, in ways of which the end user tends to be unaware. Much of the methodology and literature that deals with the storage and manipulation of graphs by computers comes in this second category. For theoretical reasons, and because of the nature of conventional computers, binary trees play a major role in a lot of this work. These are trees in which each node has, at most, two children.

5.1 Building and Saving a Tree

The program TREECON lets you construct a directed tree such as the accompanying example. The diagram is displayed on the screen. A *structure table* that represents the tree is displayed also. You can save this table on disk for later use by other programs.

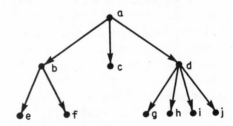

The language of family trees is applied freely to mathematical trees. Thus b, c, and d are called *children* of a, and a is called the *parent* of b, c, and d. Also, the nodes b, c, and d are called *siblings.* A node that has no children is called a *leaf* (or terminal node). The node that in family terms starts the tree is called the *root node.*

The overall action of TREECON is very simple. It begins by prompting for the root node. Suppose that you respond by typing a. The program displays a point centered horizontally near the top of the screen and labels it a. Then it prompts for the children of a. Suppose you respond by typing

```
b, c, d
```

The program displays three points labeled b, c, and d below point a. It draws lines from those points to a. Then it prompts for the children of b. For the tree in the diagram you enter the names e and f. Points appear with those labels, and lines join them to b. The program prompts for the children of e. There are none, so you just press the Enter key. You deal likewise with prompts for the children of f and c. Then the program prompts for the children of d. You enter the names g, h, i, and j. The program displays points with those labels and draws lines from them to d. It prompts for the children of g, h, i, and j in turn, and you press the Enter key in response to each of the prompts. The program then declares that the process has ended.

The actions can be restated in more general terms as follows. TREECON prompts for the root node and for the names of its children. Then TREECON deals with each child by prompting for the names of that child's children. The program deals with each child in this further generation by prompting for its children. The program continues to delve deeper and deeper into further generations until it is informed that it is prompting for the children of a node that has none. The program moves to the closest younger sibling of the childless node. If the node does not have a younger sibling the program moves to the closest antecedent that does. Whenever a sibling is found with children, the program delves into their descendants in the manner described until they have been exhausted. The process ends when the youngest child of the root node and all of its descendants (if there are any) have been processed.

The description that has just been given is essentially recursive. To deal with a node, you get its children. When there are none, you go to the closest younger sibling. When there isn't one, you go to the parent and seek the closest younger sibling.

The actions can also be formulated in terms of a stack. Begin by putting the root node onto the stack. Thereafter, pop the top node from the stack, request its children, and push the children onto the stack starting with the youngest. Then pop the top item from the stack and repeat the procedure. When a popped item has no children, pop the item that the previous popping exposed. When the stack is empty, quit.

The program TREECON uses a stack in precisely that fashion. Furthermore, it stores data that represents the complete structure of the tree in a convenient fashion, and it displays the tree in a visually convenient form. Now for the details. You can use letters or words or other strings of characters as node names without interfering with the logic of the program or harming the table that it constructs. However, because the program displays (1) the structure table that is being formed, (2) the schematic tree, and (3) the stack that is waxing and waning, any labels that are more than one letter long can obscure part of the other information on the screen.

When you respond to the initial prompt, the headings "structure table" and "stack" are displayed at the top-left and top-right of the screen, respectively. The significance of the column headings i, p, and c in the structure table will be explained shortly. The values 1, a, and 0 appear in the stack under the column headings j, name, and p. The j value is just the slot number of the other two items in the arrays that store the stacked data. The name is the name of the node. The p value identifies the parent of the node under consideration. The root node has no parent, so the value 0 is used to show that. The fact that the stack is upside down relative to the diagrams in Chapter 4 does not affect its operation; programming instructions, not gravity, are at work.

```
structure table              • a                      stack
i name p c                                         j name p
                                                   1   a   0
```

The program prompts you to press the space bar to transfer the data for a node. You respond, and the items 1, a, and 0 are transferred from the stack to the first row of the structure table. In this table, i is just the row number.

```
structure table              • a                      stack
i name p c                                         j name p
1   a   0
```

Item c is the number of children of the node. This is left blank because it is not known yet. The program prompts for the children of a. You respond by typing b, c, d. The names you just typed are transferred to the successive rows of the stack. The youngest child is pushed onto the stack first, and the eldest is at the surface (that is, lowest on the screen). The p value 1 in each row of the stack shows that the parent of all these new nodes is in row 1 of the structure table. The c value in this row is set to 3 to show the node has three children. The edges from node a to nodes b, c, and d are added to the diagram of the tree.

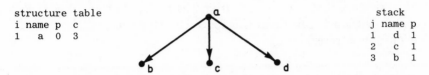

Next, you are prompted to start a "transfer and get children cycle." The name and the p value at the surface of the stack are transferred to the next available row of the structure table.

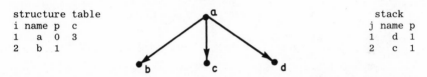

The program now prompts for the children of b, the node that was just popped and transferred. For the tree on page 194 you type the labels e and f. The new nodes are pushed onto the stack. The edges from b to the two new nodes are added to the diagram. The number of children of b appears in the structure table.

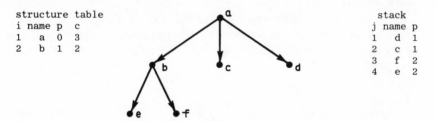

You are prompted to start another transfer-and-get-children cycle. You respond, and the node name e and its p value are popped and transferred to the structure table.

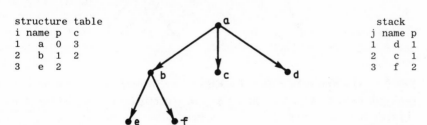

The program prompts for the children of e. There are none, and you register that by pressing the Enter key without typing a name beforehand. A 0 now appears in the appropriate position of the structure table to show that e has no children. The prompt to transfer data for a node reappears. You press the space bar, and the data for node f is popped and transferred.

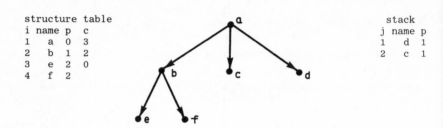

```
structure table                               stack
i name p  c                                   j name p
1  a   0  3                                    1   d   1
2  b   1  2                                    2   c   1
3  e   2  0
4  f   2
```

The next two cycles proceed in a similar manner, because f and c have no children. The responses to successive prompts put a 0 in the number-of-children column for node f. The data for node c is popped from the stack and appended to the structure table. Then the number of children is set to 0 and the data for node d is transferred, leaving the stack empty.

When the program prompts for the children of d, you type g, h, i, j. The new nodes are put into the stack. The appropriate edges are added to the diagram of the tree. The number of children of d is displayed in the structure table.

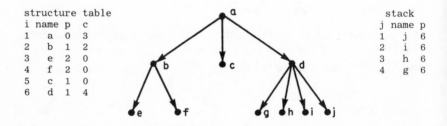

```
structure table                               stack
i name p  c                                   j name p
1  a   0  3                                    1   j   6
2  b   1  2                                    2   i   6
3  e   2  0                                    3   h   6
4  f   2  0                                    4   g   6
5  c   1  0
6  d   1  4
```

Pressing the space bar and the Enter key in response to the further prompts transfers g, h, i, and j to the structure table and sets their child counts to 0.

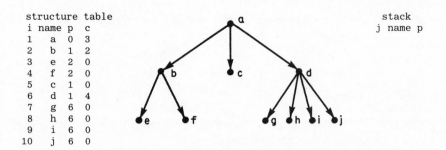

```
structure table                                          stack
i name p  c                                            j name p
1  a   0  3
2  b   1  2
3  e   2  0
4  f   2  0
5  c   1  0
6  d   1  4
7  g   6  0
8  h   6  0
9  i   6  0
10 j   6  0
```

The program now finds that the stack is empty at the point where it would prompt for the start of a transfer-and-request-children cycle. This means that every node in the tree has been accommodated either by specifying its children or by specifying the fact that it has none. Consequently, the tree is complete. The program prompts you to type S to save the structure table on disk, D to draw another tree, or Q to quit. If you type S, the program prompts for the name of the file in which the table will be saved.

The limitations of the screen restrict the number of generations and the number of collateral edges that TREECON can handle. Much larger trees can be processed by using a nonrestrictive style of display such as hierarchical indention of the names of the nodes.

Program Variables

T$(i)	The name of the ith node in the structure table
P(i)	The row number of the parent of the ith node
C(i)	The number of children of the ith node
H(i)	The generation number of the ith node (1 for the root, 2 for its children, and so on)
X(i),Y(i)	The screen coordinates of the ith node in the diagram of the tree
U$(j)	The name of the jth node in the stack
PU(j)	The row number, in the structure table, of the parent of the node in the jth row of the stack
HU(j)	The generation number of the jth node in the stack
XU(j)	The horizontal screen coordinate of the jth node in the stack
YU(j)	The corresponding vertical screen coordinate
W(j)	The width of the portion of the screen that the descendants of jth node in the stack can span

IC	The greatest number of nodes in a tree that the program will accept
I	The index of a node in the structure table
J	The row number of a node in the stack
WS	The number of characters that can be displayed across the screen, which is 80, because high resolution is used
WP	The width of the portion of the screen that the parent of the node under current attention can span
W	The width of the portion of the screen that the node under current attention and its descendants can span
H	The generation number of the nodes that the program is about to display
S$	The string that consists of the names of the children of a given node, and from which these names are removed one by one, to be pushed onto the stack
K	The index of successive children of a given node in the cycle that extracts their name from S$
KMAX	The number of children
Q	The position of the leftmost comma in S$
V$(k)	The kth name in the most recent response to the prompt for the children of a node
Y	The vertical screen coordinate of the node that will be displayed next
L	The subscript, in the stack, of the node to be displayed next in the loop that draws the branches from a node to its children which have just been named
NF$	The output file in which the structure table that represents the tree is saved
IMAX	The number of nodes in the tree

Program Notes

The internal operation of the program corresponds directly to the actions which have been described. The arrays U$, PU, XU, YU, and HU are used to stack data, pending its transfer to the structure table or other deferred use. The program first performs some simple initialization, prompts for the root node, and displays it. The index J of data on the surface of the stack is set to 1 and the appropriate data for the root node is pushed into these arrays. The program prompts for the children of successive nodes, and processes these, under the control of a WHILE . . . WEND loop that cycles until the stack is empty.

```
2010   WHILE J>0
       [transfer data for a node from the stack to the structure
       table]
2310   J=J-1
       [prompt for the children of the node that was just transferred
       and set KMAX to the number of these children]
3010     FOR K=KMAX TO 1 STEP -1
3020       J=J+1
           [transfer data for a node from the stack to the structure
           table]
3080     NEXT K
3540     FOR L=J TO J-KMAX+1 STEP -1
           [display the edge to a new node from its parent]
3610     NEXT L
3620   WEND
```

The string S$, which consists of the names of the children of the node under current attention, is unpacked into the array V$, working from left to right. Then the names are transferred from V$ to U$ in the reverse order. Thus V$ acts as a secondary stack.

Suggested Activities

1. Suppose that you are given the name of each node, together with the name of its parent, in a body of data that can be displayed as a tree. Assume that no two names are the same and that the parent of the root node is designated by a special name or by a null string. Also assume that, when a node has two or more children, the left-to-right order of the children in the tree is alphabetic. Write a program to use this set of data to form the structure table of the tree in the style used by TREECON.

2. Suppose that you are dealing with a body of data that can be displayed as a tree. Each node has a different name. Each path from the root node to a leaf (terminal node) can be represented by the string of names of the constituent nodes separated by commas or spaces. Suppose that you are given these strings for all the leaves of the tree. Write a program to convert the set of strings to the structure table of the tree. Assume alphabetic ordering of siblings.

3. Write a program that forms or loads the structure table of a tree in the style used by TREECON and displays all the ancestors of any node which you specify.

4. Extend Activity 3 to find the closest common ancestor of two nodes that you specify. [*Hint:* Start with the node that has the greater subscript in the structure table. Call this node A and the other node B. Work backward through the ancestors of A until either (1) node B is reached or (2) a node C with a lower subscript than B is reached.

```
1ØØ REM * TREECON constructs and saves a tree *
1Ø1 :
11Ø   ON ERROR GOTO 6ØØØ                                      'set error trap
12Ø   KEY OFF: SCREEN 2                                       'screen set up
13Ø   WS=8Ø                                                   'screen width
14Ø   IC=5Ø                                            'limit on # of nodes
15Ø   DIM T$(IC),P(IC), H(IC),X(IC), Y(IC), C(IC)            'for nodal data
16Ø   DIM U$(IC),PU(IC),HU(IC),XU(IC),YU(IC),V$(IC),W(IC)
999 :
1ØØØ REM * set up the tree *
1Ø1Ø   CLS                                                'clear the screen
1Ø2Ø   J=1                                                 'interim node #
1Ø3Ø   GOSUB 5ØØØ                                     'prompt for root node
1Ø4Ø   INPUT; "Name of root node: ",R$
1Ø5Ø   IF R$<>"" THEN U$(J)=R$ ELSE 7ØØØ                      'quit if null
1Ø6Ø   PU(J)=Ø                                   'interim row # of parent
1Ø7Ø   HU(J)=1                                     'interim generation #
1Ø8Ø   LOCATE 1,1                                              'headings
1Ø9Ø   PRINT " structure table" TAB(WS-6) "stack"
11ØØ   PRINT " i name p   c" TAB(WS-7) "j name p"
111Ø   LOCATE J+2,WS-8                        'display current row of stack
112Ø   PRINT USING "##    \\##"; J,U$(J),PU(J);
113Ø   XU(1)=WS*4                               'coordinates of root node
114Ø   YU(1)=1Ø
115Ø   W(1)=WS*5.5                           'width of root node section
116Ø   PSET(XU(1),YU(1))                              'display root node
117Ø   LOCATE 1+POINT(1)\8,2+POINT(Ø)\8              'and its label
118Ø   PRINT U$(1)
119Ø   I=Ø                                                 'final node #
12ØØ   P(Ø)=-1                                  'parent marker for root node
1999 :
2ØØØ REM * get the remaining nodes *
2Ø1Ø   WHILE J>Ø                        'loop until interim table is empty
2Ø15   IF I>=IC THEN P$="Too many nodes":  GOTO 62ØØ
2Ø2Ø   I=I+1                                                'final node #
2Ø3Ø   T$(I)=U$(J)                                   'transfer the label
2Ø4Ø   P(I)=PU(J)                                      'the parent row #
2Ø5Ø   H(I)=HU(J)                                 'and the generation #
2199 :
22ØØ REM * transfer a node from the stack *
221Ø   GOSUB 5ØØØ                                                'prompt
222Ø   PRINT "Start a transfer-and-get-children cycle (press space bar):";
223Ø   Z$=INPUT$(1)
2235   IF I>21 THEN 2255
224Ø   LOCATE I+2,1                                'display nodal data
225Ø   PRINT USING "##   \\## ##"; I,T$(I),P(I);    'in structure table
2255   IF J>21 THEN 228Ø
226Ø   LOCATE J+2,WS-8                            'erase from stack
227Ø   PRINT TAB(8Ø) SPC(1)
228Ø   X(I)=XU(J)                              'transfer the coordinates
229Ø   Y(I)=YU(J)
23ØØ   WP=W(J)                                 'width for subdivision
231Ø   J=J-1                               'decrement interim table pointer
2399 :
24ØØ REM * prompt for children *
241Ø   GOSUB 5ØØØ                                      'prompt for children
242Ø   PRINT "children of " T$(I);
243Ø   LINE INPUT; ": ",S$                                      'store
244Ø   K=Ø                                       'zeroize the child count
245Ø   WHILE LEN(S$)>Ø                              'loop til no residue
```

```
2460     IF LEFT$(S$,1)<>" " AND LEFT$(S$,1)<>"," THEN 2490
2470     S$=MID$(S$,2)                                        'remove space or ,
2480     GOTO 2580
2490     Q=INSTR(S$,",")                                      'find next comma
2500     IF Q>0 THEN 2550                            'jump if there is one
2505     IF K>=IC THEN P$="Too many children":  GOTO 6200
2510     K=K+1
2520     V$(K)=S$                                             'store label
2530     S$=""                                        'nullify residue
2540     GOTO 2580
2550     IF K>=IC THEN P$="Too many children":  GOTO 6200
2555     K=K+1
2560     V$(K)=LEFT$(S$,Q-1)                                  'store label
2570     S$=MID$(S$,Q+1)                              'remove from residue
2580     WEND
2590     KMAX=K                                               '# of children
2600   IF KMAX>0 THEN H=H(I)+1                                'generation #
2999 :
3000 REM * include new data in the stack and display it *
3010     FOR K=KMAX TO 1 STEP -1                      'loop through children
3015     IF J>=IC THEN P$="Stack overflow":  GOTO 6200
3020       J=J+1                                              'interim node #
3030       U$(J)=V$(K)                                        'label
3040       PU(J)=I                                            'parent node #
3050       HU(J)=H                                            'generation #
3055       IF J>21 THEN 3080
3060     LOCATE J+2,WS-8                             'display data in stack
3070     PRINT USING "##  \\##"; J,U$(J),PU(J);
3080     NEXT K
3090     C(I)=KMAX                                            '# of children
3095     IF I>21 THEN 3120
3100     LOCATE I+2,11                               'display in structure table
3110     PRINT USING "##"; C(I)
3120   IF KMAX=0 THEN 3620                           'jump if no children
3499 :
3500 REM * display new nodes and edges *
3510     W=WP/KMAX                                   'width of section for each child
3520     Y=Y(I)+32                                            'y coordinate
3530     XU(J)=X(I)-WP/2+W/2                         'x coordinate of eldest
3540     FOR L=J TO J-KMAX+1 STEP -1                 'loop through children
3550       IF L<J THEN XU(L)=XU(L+1)+W                        'x coordinate
3560       YU(L)=Y                                            'y coordinate
3570       W(L)=W                                             'width
3580       LINE (X(I),Y(I))-(XU(L),YU(L))                     'draw the edge
3582       VC=1+POINT(1)\8:  HC=2+POINT(0)\8
3584       IF VC<1 OR 24<VC OR HC<1 OR 80<=HC+LEN(U$(L)) THEN 3610
3590       LOCATE VC,HC                              'display the label
3600       PRINT U$(L)
3610     NEXT L
3620   WEND
3630   IMAX=I
3999 :
4000 REM * prompt for further action *
4010   GOSUB 5000                                            'prompt
4020   PRINT "Type S to save, D to draw another, Q to quit: ";
4030   R$=INPUT$(1)
4040   ON (1+INSTR("SsDdQq",R$))\2 GOTO 4510,1000,7000       'branch
4050   GOTO 4010                                             'retry
4499 :
4500 REM * save the data *
```

```
4510   GOSUB 5000                                              'prompt for file name
4520    INPUT; "Name of output file: ",NF$
4530     IF NF$="" THEN 4000                                    'null - try again
4540   OPEN NF$ FOR OUTPUT AS #1                                 'open the file
4550    PRINT #1, IMAX                                            '# of nodes
4560    FOR I=1 TO IMAX                                       'loop through nodes
4570     PRINT #1,T$(I)                                          'nodal data
4575     PRINT #1,P(I),C(I),X(I),Y(I)
4580    NEXT I
4590    CLOSE #1                                                       'close
4600   GOSUB 5000                                                    'prompt
4610    PRINT "Type D to draw another, Q to quit: ";
4620    IF INSTR("Qq",INPUT$(1)) THEN 7000 ELSE 1000               'branch
4999 :
5000 REM * prepare to prompt *
5010   LOCATE 25,1                                          'clear prompt line
5020    PRINT TAB(80);
5030   LOCATE 25,1                                                're-position
5040   RETURN
5999 :
6000 REM * error trap *
6010   GOSUB 5000                                           'error diagnostic
6020    PRINT "Error" ERR "in line" ERL "- press space bar"
6030    Z$=INPUT$(1)
6040   RESUME 7000
6200    GOSUB 5000
6210    PRINT P$ ". Press space bar to go on.";
6220    Z$=INPUT$(1)
6230    GOTO 4000
6999 :
7000 REM * quit *
7010   CLOSE                                                 'close all files
7020   LOCATE 1,1                                             'for Ok message
7030   ON ERROR GOTO 0                                  'reset error trap address
7040   END                                                            'quit
```

In case 1, B is an ancestor of A. In case 2, work backward through the ancestors of B until node C, or a node with a lower subscript, is reached. Continue to alternate in this way until one of the nodes in the pair under consideration is found to be an ancestor of the other.]

5. Extend Activity 4 to find the actual path between two nodes that is specified in response to a prompt. (*Hint:* Push the ancestors of the two starting nodes onto two stacks.)

6. Alter TREECON to permit the input of data that defines much larger trees by suppressing the displayed diagram.

7. Expand the program TREECON to allow specification of the lengths of the edges of a tree. Expand the structure table to include the lengths. Extend Activity 5 to find the distance between two nodes by adding the lengths of the edges that lead to the common ancestor.

8. Write a program to merge two trees that have the same root node. Assume alphabetic ordering of siblings. For example, the result of merging trees (*a*) and (*b*) is tree (*c*).

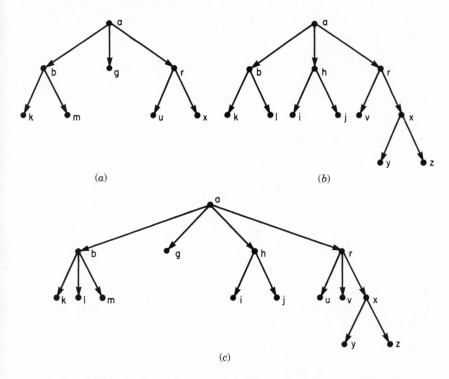

(*a*) (*b*)

(*c*)

9. Expand Activity 8 to form other combinations of trees, for example (*a*) the intersection of two trees with the same root node, (*b*) the result of deleting from tree A all nodes and any descendants they may have which match terminal nodes of tree B, (*c*) the result of deleting from tree A all nodes that match terminal nodes of B together with any descendants they may have and any ancestors that have no other terminal descendants.

5.2 Finding All Possible Paths

The TREECON program in Section 5.1 lets you create a structure table that defines a tree in a simple fashion. The program TRAVALL, which is described in the present section, reloads the structure table, displays a diagram of the tree, and lists the paths from the root node to all the terminal nodes (leaves). In the course of identifying a path through the tree, the program intensifies the constituent edges in the diagram. This helps you see how the program keeps track of where each path diverges from its predecessor in the systematic traversal process which is followed. TRAVALL shows how the structure table can be expanded mechanically to help find siblings, cousins, and other relatives of a node very quickly. It also provides a further example of the use of stacks when dealing with trees.

The program begins by prompting for the name of the file that contains the structure table. Then the program reads the file and displays the tree and an expanded form of the table.

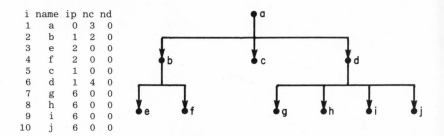

i	name	ip	nc	nd
1	a	0	3	0
2	b	1	2	0
3	e	2	0	0
4	f	2	0	0
5	c	1	0	0
6	d	1	4	0
7	g	6	0	0
8	h	6	0	0
9	i	6	0	0
10	j	6	0	0

The successive rows of the table correspond to the nodes in the order explained in Section 5.1. This is called top-to-bottom, left-to-right order. It is the age-old order of dynastic succession. Each row of the table contains (1) the row number, (2) the name of the node to which the row refers, (3) the row number of its parent, (4) the number of children, and (5) the total number of descendants. The columns are headed i, name, ip, nc, and nd, respectively. The nd column contains zeros at the outset.

By using just horizontal and vertical lines in the displayed tree, the edges are of uniform intensity, initially, and also can be intensified uniformly. This would be much more difficult to achieve with lines that sloped by different amounts, in the style of the tree on page 194.

The program now constructs the total number of descendants of each node to facilitate the later operations. It prompts you to press the space bar to start computing the numbers of descendants.

You respond, and the program starts with the final row of the table, by prompting you to allow for node 10 in the nd value of its parent (node 6). You press the space bar, and the nd value in row 6 is increased by 1. Attention moves back one row, and you are prompted to allow for node 9 in the nd value of its parent (node 6). You respond, and the nd value in row 6 is increased by 1 again. You continue in this way to allow for nodes 8 and 7 in the nd values of their parent. This is node 6, and its nd value is increased to 3 and 4, in turn. Then you allow for node 6 in the nd value of its parent, that is, node 1. This is increased from 0 to 5 to allow for node d and its four descendants. Thus, in each cycle of the process to count descendants, the nd value of the parent of the node under attention is increased by 1 more than the nd value of that node. Continuing in this way, consideration of node 5 increases the nd value of node 1 to 6. Consideration of nodes 4 and 3 increase the nd value of node 2 to 1 and then to 2. Consideration of node 2 increases the nd value of node 1 to 9.

```
 i  name  ip  nc  nd
 1    a    0   3   9
 2    b    1   2   2
 3    e    2   0   0
 4    f    2   0   0
 5    c    1   0   0
 6    d    1   4   4
 7    g    6   0   0
 8    h    6   0   0
 9    i    6   0   0
10    j    6   0   0
```

The program now starts to construct the paths from the root node to the terminal nodes by following the line of eldest children from generation to generation. Then it backtracks repeatedly to the closest ancestor who has a younger sibling and follows the lines of descent from there. Pressing the space bar twice puts the headings of four columns of a stack at the top-right corner of the screen and pushes the relevant data for the root node into them.

```
j  name  i  nc  n
1   a    1   3   0
```

The successive items are, respectively, the row number of the data in the stack, the label of the node, the row number of the node in the structure table, the number of the node's children, and a "child number" that is 0 initially. The program prompts you to include node a in the current path. The caption "Current path:" appears at the bottom of the screen. You press the space bar, and the bottom line changes to "Current path: a". Now, because the child number (0) in the top row of the stack is less than the number of children (3), the program recognizes that the node (a) has at least one child that can be included in a path to one or more terminal nodes. The prompt changes to:

```
Node a has a child. Press space bar to find it
```

You respond. The child number of a is increased to 1 in the n column of the stack. Then the data for the first child of a is pushed onto the stack.

j	name	i	nc	n
1	a	1	3	1
2	b	2	2	0

The edge from node a to node b is intensified in the diagram. You are prompted to press the space bar to include node b in the current path. The bottom line changes to "Current path: ab".

You are prompted that node b has a child, and you press the space bar to find it. The data for the first child of b is pushed onto the stack after the child number of b has been increased to 1.

j	name	i	nc	n
1	a	1	3	1
2	b	2	2	1
3	e	3	0	0

The edge from node b to node e is intensified in the diagram. You are prompted to include node e in the path, and the bottom line changes to: "Current path: abe". Node e is terminal. The program recognizes this from the fact that the number of its children is 0. The program prompts:

```
Node e is terminal. Press space bar to store the path
```

You respond, and the path abe is stored in a list of paths. Then the program prompts you to backtrack to the parent (node b). You respond. The node e is deleted from the current path in the bottom line of the display. The data about node e is erased from the surface of the stack, which exposes the data about its parent, node b. The edge from b to e in the diagram is deemphasized. You are prompted that node b has another child, and you press the space bar to find it.

The program can tell that b has another child, because the child number (in the column headed n) is less than the total number of children of b (in the column headed nc) in the surface row of the stack. The data about the next child of b in the structure table is found very simply. The row number of b is just increased by the total number of descendants of b plus 1. Thus, the data about the sibling is found in row 3 (the row number of b) plus 0 (the number of descendants of b) plus 1; that is row 4. The data in the stack changes to:

j	name	i	nc	n
1	a	1	3	1
2	b	2	2	2
3	f	4	0	0

The edge from b to f is emphasized in the diagram of the tree. The program prompts in turn (1) to include node f in the current path, (2) to store the path, because node f is terminal, and (3) to backtrack to the parent of f. This is done simply by going back to the preceding row of the stack.

On backtracking to b this time, the program finds that the child number (that is, the n value 2 in the stack) is no longer less than the child count (that is, the nc value 2 in the stack). It prompts:

```
Node b has no more children. Press space bar to backtrack
```

You respond. The stack at this point contains just the one row of data for node a. The program recognizes that a has another child because the n value, 1, is less than the nc value, 3. The pointer i that was just erased from the stack shows that data for the node from which the program just backtracked is in row 2 of the structure table. The number of descendants of this node, given by the nd value, is 2. Consequently, the data for its younger sibling is in row $2 + 2 + 1$ of the structure table; that is row 5. It is node c.

The child count of node a is increased by 1 in the surface row of the stack. The data about node c is pushed onto the stack.

j	name	i	nc	n
1	a	1	3	2
2	c	5	0	0

The edge from node a to node c is included in the current path, which becomes ac. The program finds that node c is terminal, and it backtracks. The stack contracts to:

j	name	i	nc	n
1	a	1	3	2

Action proceeds in the mold that has been established. The data for node d is pushed onto the stack and node d is incorporated in the current path. The program finds that this has a child and identifies it as g. It is incorporated in the current path. Its data is pushed onto the stack. The program finds it is terminal, backtracks, and then descends to h. Similar sequences of actions backtrack and then descend to node i and then j. The stack contains the following data at this point

j	name	i	nc	n
1	a	1	3	3
2	d	6	4	4
3	j	10	0	0

Backtracking leads to node d, which has no more children, and then to node a, which has no more children either. Therefore, all paths have been traced.

Program Variables

IC	The maximum number of nodes in a tree that the program can handle
IMAX	The number of nodes in the tree under consideration
I	The index of a node in the structure table. This takes the values $1, \ldots,$ IMAX for the nodes in top-to-bottom, left-to-right order
T\$(i)	The label of the ith node
P(i)	The row number, in the structure table, of the parent of the ith node
C(i)	The number of children of the ith node
ND(i)	The total number of descendants of the ith node
X(i),Y(i)	The screen coordinates of the ith node in the diagram
JC	The maximum length of a path allowed by the program (actually the stack capacity)
J	The subscript of data in the stack
IS(j)	The row number, in the structure table, of the data about the node to which the jth row of the stack refers
N(j)	The child number in the jth row of the stack (that is, 1, when the first child of the node described in the stack is in the current path, 2 when the second of these children is there, and so on)
KC	The maximum number of paths that can be stored for later redisplay
PATH\$(k)	The string of nodes that constitutes the kth path
WS	The number of characters that can be displayed across the screen, which is 80, because high-resolution graphics are used
TRSW	The edge intensity switch that is 0 to draw an edge initially, 1 to emphasize the edge, and 2 to deemphasize the edge
MS\$	The text of the current prompt
P\$	The string "Press space bar to" that appears in most of the prompts
NF\$	The name of the input file
K	The subscript of the current path in the array of paths that is formed
NEWP\$	The string of node names in the portion of the current path that has been formed so far

FWD   The switch that is on when the current path is being extended by the inclusion of further descendants

BWD   The switch that is on when backtracking is in progress

ICC   The row number, in the structure table, of the node whose data has just been removed from the stack

KK   The subscript of the path in the loop that redisplays the paths

YMID   The vertical screen coordinate of the midpoint of the edge that is being displayed, emphasized, or deemphasized

Program Notes

The program illustrates the tidy manner in which trees are processed systematically by WHILE . . . WEND loops and the ease of finding nodes that are related to each other. After some trivial initialization, the program prompts for the input file, reads it, displays the structure table, and displays the tree. It uses the subroutine at line 3000 to draw each edge.

The number of descendants of each of the nodes is found by the statements

```
480   FOR I=IMAX TO 2 STEP -1
510     ND(P(I))=ND(P(I))+ND(I)+1
540   NEXT I
```

The pointers J, I, and K are initialized to 1, 1, and 0, respectively. NEWP\$ is nullified. Then the traversal of all the paths is performed by the loop

```
1010 WHILE J>0
. . .
1610 WEND
```

Within this loop, the descent to a terminal node and the incorporation of the relevant nodes in the current path are performed by the following statements (the in-line REMs are adapted from those in the program listing.

```
1020  FWD=-1                            'descendants are possible
1110  WHILE FWD                          'loop while this is so
1160    NEWP$=NEWP$+T$(I)                     'append node I
1210    IF C(I)=0 THEN FWD=0: GOTO 1310
                          'turn off descent switch if childless
1240    N(J)=1                     'set child number for first born
1270    IS(J)=I               'stack pointer to structure table
1280    I=I+1                            'point to first born
1300    J=J+1                        'increase stack pointer
1310  WEND
```

Following the recognition of a terminal node, the path is stored by

```
1440    PATH$ (K) =NEWP$
```

Then backtracking is performed by the statements

```
1460    BWD=-1                                      'backtracking is needed
1470    WHILE BWD AND J>0              'backtrack while needed and possible
1490       NEWP$=LEFT$ (NEWP$, LEN (NEWP$) -1)         'remove rightmost node
1540       ICC=I                          'preserve pointer to structure table
1550       J=J-1                               'decrease the stack pointer
1560       IF J=0 THEN 1800                      'jump if stack is empty
1570       I=IS (J)                           'pointer for parent node
1580       IF C (I)>N (J)  THEN BWD=0:  GOTO 1610
                                            'stop backtracking if more children
1610    WEND
```

Suggested Activities

1. Write a program that lists the paths to the root node from every terminal node of a tree. Use a structure table formed by TREECON, but do not use a stack. [*Hint:* Work through the successive rows of the structure table. For each childless node (nc value is 0), work backward through its ancestors (use the ip values).]

2. Expand Activity 1 as follows. Store the paths in an array and associate the subscript of the first path that contains a given node with the data for that node in the structure table. In working backward from a childless node, when a node is reached that has been included in an earlier path, copy the relevant portion of that path to complete the path that is being formed. Represent a path by the concatenation of the names of its constituent nodes.

3. Suppose that you have a modified structure table of a tree in which successive rows represent the nodes in top-to-bottom, left-to-right order and in which each row contains just the name of the node and the total number of the node's descendants. Write a program to expand the data for each node to include the pointer to its parent. (*Hint:* To find the parent of a node, set a "minimum descendant number" to the number of descendants of that node. Work backward row by row while adding 1 to that number in each cycle until you reach a node with a total number of descendants that equals or exceeds the value that the minimum descendant number has reached. This node is the parent of the node from which you started.)

4. Expand Activity 3 to find the number of children of each node. (*Hint:* A node with 0 descendants has no children. Otherwise, the

```
100 REM * TRAVALL - traverses all paths from root node to terminal nodes *
101 :
110   ON ERROR GOTO 5000                                  'set error trap
120   IC=50                                               'limit on # ofnodes
130   JC=50                                               'stack capacity
140   KC=50                                               'limit on # of paths
150   DIM T$(IC),P(IC),C(IC),ND(IC),X(IC),Y(IC)            'for nodal data
160   DIM IS(JC),N(JC),PATH$(KC)                    'for stacked data & paths
170   KEY OFF: SCREEN 2                                    'screen set up
180   WS=80                                                'screen width
190   P$="Press space bar to "                             'for prompts
199 :
200 REM * set up the tree *
210   CLS
220   TRSW=0                                        'edge intensity switch
230   GOSUB 4500
240   INPUT; "name of input file: ",NF$                         'prompt
250   OPEN NF$ FOR INPUT AS #1                           'open input file
260   INPUT #1, IMAX                                     '# of nodes
270   IF IMAX>IC THEN ERROR 251                   'error if too many nodes
280   FOR I=1 TO IMAX                                   'loop through nodes
290     INPUT #1, T$(I)                                       'label
300     INPUT #1, P(I),C(I),X(I),Y(I)      'parent node, # of children, coords
302     X(I)=X(I)+30                               'allow for wider table
304     VC=1+Y(I)\8:  HC=2+X(I)\8
306     IF VC<1 OR 24<VC OR HC<1 OR 80<HC+LEN(T$(I)) THEN 330
310     LOCATE VC,HC
320     PRINT T$(I);
330     GOSUB 3000                                    'display the edge
350   NEXT I
360   CLOSE #1                                        'close the file
399 :
400 REM * display structure table *
410   LOCATE 1,1
420   PRINT " i name ip nc nd"                   'nodal data table heading
425   ERASE ND: DIM ND(IC)
430   FOR I=1 TO IMAX                             'display nodal data
440     IF I<21 THEN PRINT USING "##   !  ## ## ##"; I,T$(I),P(I),C(I),0
450   NEXT I
460   MS$=P$+"start computing number of descendants"            'prompt
470   GOSUB 4000
480   FOR I=IMAX TO 2 STEP -1                     'compute # of descendants
490     MS$=P$+"allow for node"+STR$(I)+
              " in nd value of its parent (node"+STR$(P(I))+")"
500     GOSUB 4000
510     ND(P(I))=ND(P(I))+ND(I)+1                 'increase # of descendants
515     VC=P(I)+1:  IF VC>24 THEN 540
520     LOCATE VC,15
530     PRINT USING "##";ND(P(I))                        'display new #
540   NEXT I
599 :
600 REM * start the traversal *
610   MS$=P$+"start the first path"               'prompt to start 1st path
620   GOSUB 4000
630   LOCATE 1,WS-13                                 'heading of stack
640   PRINT "j name i nc  n"
650   MS$=P$+"start with the root node"                      'prompt
660   GOSUB 4000
670   J=1                                         'initialize stack pointer
680   I=1                                                'head node
```

```
 690   NEWP$=""                                                    'path string
 700   K=0                                                         'and path #
 710   LOCATE 25,1
 720    PRINT "Current path: ";                                    'caption
 730   ERASE N: DIM N(JC)
 999 :
1000 REM * main cycle *
1010   WHILE J>0                                       'loop until stack is empty
1020    FWD=-1                                         'descendant switch on
1030    IF K>=KC THEN ERROR 253                              'too many paths
1040    K=K+1                                                 'increase path #
1099 :
1100 REM * find descendants *
1110   WHILE FWD                                  'loop until terminal node
1115    IF J>23 THEN 1140
1120    LOCATE J+1,WS-14                                  'display stacked data
1130     PRINT USING "##   ! ## ## ##"; J,T$(I),I,C(I),N(J)
1140    MS$="Press space bar to include node "+T$(I)+" in current path"
1150     GOSUB 4000                                            'prompt
1160    NEWP$=NEWP$+T$(I)                           'append node to current path
1170    TRSW=1                                             'intensify the
1180     GOSUB 3000                                        'displayed edge
1190    LOCATE 25,15                                  'display the lengthened path
1200     PRINT NEWP$ TAB(79);
1210    IF C(I)=0 THEN FWD=0: GOTO 1310               'jump if path ended
1220    MS$="Node "+T$(I)+" has a child. "+P$+"to find it"
1230     GOSUB 4000                                            'prompt
1240    N(J)=1                                       'child count
1250    IF J<21 THEN LOCATE J+1,WS-1 ELSE 1270
1260     PRINT USING "##"; N(J)                                   'display
1270    IS(J)=I                                     'stack the node #
1280    I=I+1                                        'successor node #
1290    IF J>=JC THEN ERROR 252
1300    J=J+1                                   'increase the stack pointer
1310   WEND
1399 :
1400 REM * backtrack *
1410   MS$="Node "+T$(I)+" is terminal. "+P$+"store the path"      'terminal node
1420    GOSUB 4000
1430    TRSW=2                                            'de-emphasize the edge
1435    GOSUB 3000
1440    PATH$(K)=NEWP$                                      'store the path
1450    MS$=P$+"backtrack to parent (node "+T$(P(I))+")"           'for prompt
1460    BWD=-1                                             'bactrack switch on
1470    WHILE BWD AND J>0            'loop to new descent path or til stack empty
1480     GOSUB 4000                                           'prompt
1490     NEWP$=LEFT$(NEWP$,LEN(NEWP$)-1)             'delete end node from path
1500     LOCATE 25,15                                       'redisplay
1510      PRINT NEWP$ TAB(79);
1520     IF J<20 THEN LOCATE J+1,WS-14 ELSE 1540    'erase displayed stack line
1530      PRINT TAB(80) SPC(1)
1540     ICC=I                               'node that may have sibling
1550     J=J-1                               'decrease stack pointer
1560     IF J=0 THEN 1800                    'jump if empty
1570     I=IS(J)                             'pop node #
1580     IF C(I)>N(J) THEN BWD=0: GOTO 1610   'turn backtrack switch off & jump
1590     MS$="Node "+T$(I)+
                    " has no more children. "+P$+"backtrack"       'prompt
1600    GOSUB 3000                                      'de-emphasize edge
1610   WEND
```

```
1699 :
17ØØ REM * use next child *
171Ø   MS$="Node "+T$(I)+" has another child. "+P$+"find it"              'prompt
172Ø     GOSUB 4ØØØ
173Ø     N(J)=N(J)+1                                          'increase child count
174Ø     IF J<21 THEN LOCATE J+1,WS-1 ELSE 176Ø                          'redisplay
175Ø       PRINT USING "##"; N(J)
176Ø     IF J>=JC THEN ERROR 252                                    'stack overflow
177Ø     J=J+1                                              'increase stack pointer
178Ø     I=ICC+ND(ICC)+1                                       'node # of sibling
179Ø     TRSW=1                                                'emphasize the edge
18ØØ     GOSUB 3ØØØ
181Ø   WEND
182Ø   LOCATE 25,1
183Ø     PRINT TAB(79);
1999 :
2ØØØ REM * prompt for further action *
2Ø1Ø   LOCATE 23,1                                                       'prompt
2Ø2Ø     PRINT "Type A for another, D to display paths, R to repeat, Q to quit"
               TAB(79)
2Ø3Ø     R$=INPUT$(1)
2Ø4Ø     ON ((1+INSTR("AaDdRrQq",R$))\2) GOTO 2ØØ,25ØØ,4ØØ,6ØØØ           'branch
2Ø5Ø     GOTO 2Ø1Ø
2499 :
25ØØ REM * display the paths *
251Ø   FOR KK=1 TO K
252Ø     MS$=P$+"display path"+STR$(KK)
253Ø     GOSUB 4ØØØ
254Ø     LOCATE 25,1
255Ø       PRINT "Path" KK "consists of nodes " PATH$(KK) TAB(79);
256Ø   NEXT KK
257Ø   LOCATE 25,1
258Ø     PRINT TAB(79)
259Ø     GOTO 2ØØØ
2999 :
3ØØØ REM * display an edge *
3Ø1Ø   IF I=1 THEN RETURN
3Ø2Ø     IP=P(I)                                                 'parent node #
3Ø3Ø     YMID=(Y(I)+Y(IP))/2                                  'y coord of mid-point
3Ø4Ø   ON TRSW GOTO 32ØØ,33ØØ                            'branch on intensity switch
31ØØ     LINE (X(IP),Y(IP))-(X(IP),YMID)                     'upper half vertical
311Ø     LINE (X(IP),YMID)-(X(I),YMID)                               'horizontal
312Ø     LINE (X(I),YMID)-(X(I),Y(I))                         'lower half vertical
313Ø     RETURN
32ØØ     LINE (X(IP)-1,Y(IP))-(X(IP)-1,YMID)          'intensify upper half vertical
321Ø     LINE (X(IP)+1,Y(IP))-(X(IP)+1,YMID)
322Ø     LINE (X(IP),YMID-1)-(X(I),YMID-1)                           'horizontal
323Ø     LINE (X(I)-1,YMID)-(X(I)-1,Y(I))                     'lower half vertical
324Ø     LINE (X(I)+1,YMID)-(X(I)+1,Y(I))
325Ø     RETURN
33ØØ     LINE (X(IP)-1,Y(IP))-(X(IP)-1,YMID),Ø  'de-emphasize upper half vertical
331Ø     LINE (X(IP)+1,Y(IP))-(X(IP)+1,YMID),Ø
332Ø     LINE (X(IP),YMID-1)-(X(I),YMID-1),Ø                         'horizontal
333Ø     LINE (X(I)-1,YMID)-(X(I)-1,Y(I)),Ø                   'lower half vertical
334Ø     LINE (X(I)+1,YMID)-(X(I)+1,Y(I)),Ø
335Ø     RETURN
3999 :
4ØØØ REM * prompt for key press *
4Ø1Ø   LOCATE 23,1                                                 'prompt line
4Ø2Ø   PRINT MS$ "." TAB(79)
```

```
4030   Z$=INPUT$(1)                                    'wait for key press
4040   RETURN
4500 REM * prepare for a prompt *
4510   LOCATE 23,1                                     'clear prompt line
4520   PRINT SPACE$(79);
4530   LOCATE 23,1
4540   RETURN
4999 :
5000 REM * error trap *
5010   MS$="Error"+STR$(ERR)+" in line"+STR$(ERL)      'general diagnostic'
5020   IF ERR=251 THEN MS$="Too many nodes"            'special cases
5030   IF ERR=251 THEN MS$="Stack overflow"
5040   IF ERR=253 THEN MS$="Too many paths"
5050   MS$=MS$+". "+P$+"go on"
5060   GOSUB 4000                                      'prompt
5070   RESUME 6000
5999 :
6000 REM * quit *
6010   CLOSE                                           'close all files
6020   LOCATE 20,1                                     'for Ok message
6030   ON ERROR GOTO 0                                 'reset error trap address
6040   END
```

row that describes the node is followed by the row that describes the node's eldest child. Proceed from each child to its younger sibling by increasing the row number by the number of descendants of the child until the result exceeds the sum of the row number of the parent and the total number of descendants of the parent.)

5. Write a program that stores the structure table of a tree and reverses the order of the children of each node. Assume the structure table includes the pointer to the parent and the number of children and descendants of each node.

6. Write a program that converts the structure table of a tree into the structure table of a tree that consists of the same set of nodes and edges but has a different root node that is specified in response to a prompt.

7. If x is the name of a node, let P.x denote the parent, F.x the eldest child, L.x the youngest child, E.x the elder sibling, and Y.x the younger sibling of x, respectively, when they exist, and a null string otherwise. Let A.x, C.x, D.x, G.x, H.x, and S.x denote the total number of ancestors of x, the number of children of x, the number of descendants of x, the generation of x, the number of nodes to the leaf farthest from x, and the number of nodes to the leaf closest to x, respectively. Write a program to find, from a structure table, the value of any expression that consists of a valid sequence of the operators P, . . . , S, followed by the name of a node.

8. Write a program that (1) stores or forms a structure table, (2) positions a tree-climbing cursor at the root node, and (3) constructs output that is defined by tree-climbing expressions as follows. Each expression consists of a succession of operators that either (1) move the cursor or (2) write information about the node under current consideration into the output. Use the operators P, F, L, E, Y, and R to move the cursor from one node to its parent, eldest child, youngest child, elder sibling, younger sibling, and the root node, respectively. Let the operator V write the name of the node where the cursor is positioned into the output when the expression is valid. If Z stands for any of the tree-climbing operators and n stands for a number, let nZ denote n applications of Z in succession. Let A, C, D, G, H, and S denote the numbers associated with the node under attention that correspond to the definitions of A.x, . . . , S.x in Activity 7.

9. Extend Activity 8 to trees in which each node is a multifield record. Let Vk write the contents of the kth field of the node into the output.

5.3 Forming a Binary Tree

A vast amount of computing activity deals with *binary trees*. Each node in a binary tree has either two descendants, one descendant, or no descendants. Binary trees play a special role in computing, both because of theoretical considerations and because of certain features of conventional computer hardware. The term *general tree* is used for a tree in which the number of children of each node is unrestricted.

The program BINCON demonstrates the conventions that are commonly used to represent a general tree by a binary tree. BINCON also demonstrates the algorithm that performs the conversion and saves a table on disk that can be read by other programs that operate on the binary tree.

BINCON begins by displaying the prompt:

```
(P)reset data, (D)isk file, or (Q)uit:
```

If you type a P for preset data, the program displays the general tree:

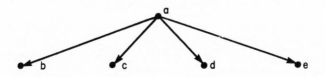

Pressing the space bar several times converts this tree, step by step, into the following "rectified" form.

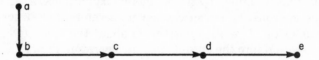

Thus, the parent node is moved into vertical alignment with the oldest child. The edge from the parent to the oldest child is drawn vertically. A horizontal edge joins each child to its younger sibling. The rectification actually proceeds from right to left on the screen to let you see one edge redrawn at a time.

After the tree has been rectified, the program draws a line beneath the rectified tree and redraws the tree, lower down the screen, in the standard format used for a binary tree:

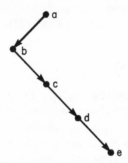

Thus, a vertical edge in the rectified tree is drawn as an edge that slopes down and to the left. A horizontal edge in the rectified tree is drawn as an edge that slopes down and to the right. You are prompted to press the space bar to make the program draw each edge. After drawing the binary tree, the program prompts:

```
(D)isplay links,  (S)ave, (R)epeat, or  (M)ain menu:
```

If you type D, the following table is displayed in the top-left corner of the screen:

```
parent left right
   a      b     /
   b      /     c
   c      /     d
   d      /     e
   e      /     /
```

Thus, the nodes are listed in top-to-bottom, left-to-right order in the binary tree. The names of the left and right children are displayed alongside the name of each parent. A slash denotes the absence of a child. After displaying this table, the program repeats the prompt to type D, S, R, or M.

If you type S, the program prompts for the name of the output file. You respond. The program saves the tabular depiction of the binary tree and redisplays the prompt to type D, S, R, or M. If you type R, the screen clears, the original tree is redisplayed, and you press the space bar to rectify the tree and redraw the binary tree step by step. If you type M, the main menu reappears.

Type P again to display another tree that is defined by preset data. The following tree is displayed:

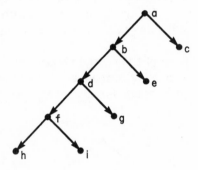

This is converted to the rectified tree:

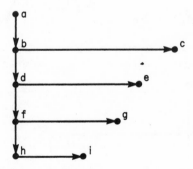

Then the program draws the horizontal dividing line and displays the sloping binary tree:

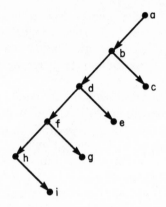

Note that, although the original tree happened to be a binary instance of a general tree, its standard binary representation is different. This example, incidentally, illustrates another feature of the program, which is to scroll the display vertically when it reaches the bottom of the screen.

After the binary form of the second preset tree has been displayed, the prompt to type D, S, R, or M reappears. If you type M for the main menu and then type P again for another preset tree, the program displays the tree:

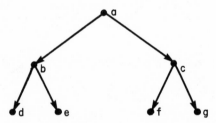

Pressing the space bar converts this, step by step, to the rectified form:

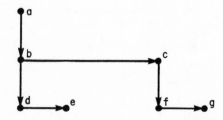

The horizontal dividing line is drawn beneath the rectified tree, and then the sloping binary tree is drawn.

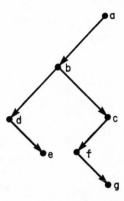

The edges ab, bd, and bc are 1½ times the length of the lower edges to prevent nodes e and f from coinciding on the screen. The program actually proceeds to display the tree, generation by generation, with edges that have the default length until the edge de has been drawn.

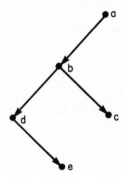

The program recognizes that continuing to use edges of uniform length will bring node f into coincidence with e. It prompts:

```
Press space bar to redraw tree - nodes will coincide otherwise
```

After the third preset tree has been displayed in binary form, the program provides two more preset trees that contain 10 and 29 nodes, respectively. Then typing M and P again simply restarts the cycle of five preset trees. If you type D in response to the main menu, the program prompts:

```
Name of input file (default is TREEDATA):
```

You can enter the name of a file created by the program TREECON. Pressing the Enter key makes this default to TREEDATA, which is the default name of the output file in TREECON. If this represents the tree that illustrated Sections 5.1 and 5.2, the BINCON program displays the general tree:

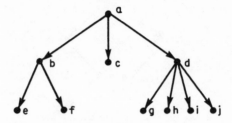

Pressing the space bar repeatedly converts this to the rectified form:

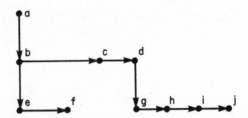

The horizontal dividing line is drawn, and then the sloping form of the binary tree is drawn:

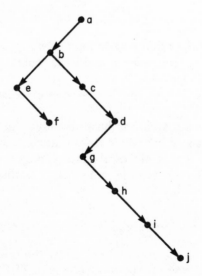

The algorithmic basis of the transformation is straightforward. The top-to-bottom, left-to-right order of the nodes remains unchanged. The input tree is specified by a structure table in the manner described in Sections 5.1 and 5.2. Within this table, each node is represented by

1. Its name

2. The number of its children

3. Its coordinates in the graphic display

4. The node number of its parent

The connectivity is defined unambiguously by giving either the set of values 2 or the set of values 4. The program TREECON records all the items 1 to 4 for each node, however, for convenience. We denote items 2 and 4 for the ith node by $c(i)$ and $p(i)$, respectively, in the discussion that follows. We denote the total number of descendants of the ith node by $d(i)$.

After the tables that define a general tree have been read from preset data or from an input file, the descendant counts $d(i)$ are constructed in the manner described in Section 5.2. The program simply initializes all the $d(i)$ to zero and then works backward through the nodes and increases $d(p(i))$ by $d(i) + 1$ for i taking values $n, n - 1, \ldots, 2$, where n is the number of nodes in the tree.

The connectivity of the binary tree is represented by the set of values $b(i,1)$ and $b(i,2)$, which are the node numbers of the left and right children of the ith node, respectively. If the ith node of the binary tree does not have a left child, $b(i,1)$ is 0. Correspondingly, $b(i,2)$ is 0 when the ith node does not have a right child. The transformation thus constructs the following tables for the four trees that have been discussed.

Preset tree 1

Node number	1	2	3	4	5
Node name	a	b	c	d	e
b(i,1)	2	0	0	0	0
b(i,2)	0	3	4	5	0

Preset tree 2

Node number	1	2	3	4	5	6	7	8	9
Node name	a	b	d	f	h	i	g	e	c
b(i,1)	2	3	4	5	0	0	0	0	0
b(i,2)	0	9	8	7	6	0	0	0	0

Preset tree 3

Node number	1	2	3	4	5	6	7
Node name	a	b	d	e	c	f	g
b(i,1)	2	3	0	0	6	0	0
b(i,2)	0	5	4	0	0	7	0

Tree used in Sections 5.1 and 5.2

Node number	1	2	3	4	5	6	7	8	9	10
Node name	a	b	e	f	c	d	g	h	i	j
b(i,1)	2	3	0	0	0	7	0	0	0	0
b(i,2)	0	5	4	0	6	0	8	9	10	0

The table of b values is formed by working through the structure table of the input tree. If $c(i)$ is greater than 0, node i has one or more children. The eldest is node $i+1$. Consequently, b(i,1) is set to $i+1$. Furthermore, if i' is the node number of any child of node i except the youngest, $b(i',2) = i' + d(i') + 1$.

The array of b values defines the connectivity of the rectified and sloping binary trees. To draw the sloping tree compactly, but without letting nodes coincide on the screen, three more arrays are constructed. Let g(i) denote the generation of the ith node in the binary tree. Let l(g) denote the horizontal distance between a node of generation g and either of its children. Let iq(q) denote the number (in the top-to-bottom, left-to-right traversal) of the node that comes qth in the traversal that proceeds completely from left to right through one generation of the tree before going on to the next.

We start by setting the edge lengths l(g) to a uniform value and proceed to draw the edges from successive nodes working from left to right through each generation in turn. Before drawing the edge to the left child of a node, the program tests if the preceding node of the generation under consideration has a right child. This condition can be written:

```
b(iq(q),1)>0    b(iq(q-1),2)>0    g(iq(q))=g(iq(q-1))
```

If this condition has occurred, the l values for the generations 1, . . . , $g-1$ are increased by a factor of $\frac{3}{2}$, the partial tree is erased from the screen, and the cycle of displaying successive nodes and edges is recommenced.

Program Variables

OP	The numerical code 1, 2, . . . corresponding to the response to the main prompt
LPM	The number of trees defined by preset data
LP	The index 1, 2, . . . for the successive trees
NF$	The file of data that defines the input tree when it is read from disk
DF$	The default file that is used to define the tree
IMAX	The number of nodes in the tree under current attention
I	The identification number of a node, in top-to-bottom, left-to-right enumeration, that also serves as the subscript of the arrays that define the input tree and its binary representation
T$(i)	The name of the ith node
P(i)	The I value of the parent of the ith node in the input tree
C(i)	The number of children of the ith node

X(i),Y(i)	The screen coordinates of the ith node
D(i)	The total number of descendants of the ith node
B(i,1)	The node number of the left child of the ith node in the binary tree
B(i,2)	The node number of the right child of the ith node
XB(i),YB(i)	The screen coordinates of the ith node in the binary tree
G(I)	The generation of the ith node in the binary tree
GMAX	The maximum value of the G(i), that is, the height of the binary tree
IQ(k)	The I value of the kth node in the generation-by-generation, left-to-right ordering of the binary tree
IL	The predecessor of node I in the sequence represented by the IQ array
Q	The subscript of the IQ array
NC	The number of children of the node under current attention in the input tree
XP,YP	The screen coordinates of the node under current attention
TC$(j)	The name of the jth child of the node under current attention
XC(j),Y(j)	The screen coordinates of the jth child
JC	The subscript 1, 2, . . . , for the successive children of the node under current attention
IC	The node number in the entire tree of the child under current attention
XC,YC	The coordinates of the eldest child
IP	The node number of the parent that is being aligned vertically with its oldest child
IG	The node number of the grandparent of node IC
XG,YG	The coordinates of node IG
VAC	The code 1, . . . , 4 for the respective cases (1) the root node is being aligned with its eldest child, (2) an eldest child is being aligned with its eldest child, (3) a middle child is being aligned with its eldest child, (4) a youngest child is being aligned with its eldest child
YDIV	The vertical coordinate of the horizontal line drawn below the general tree
M	The index that takes the values 1 and 2 for the left and right children, respectively, of a node in the binary tree in some loops that deal with the tree
K	The node number of the child under current attention in the binary tree

```
100 REM * BINCON converts a general tree to the corresponding binary form *
101 :
110  ON ERROR GOTO 8000                                          'error trap
120  SCREEN 2                                               'set up the screen
130   CLS
140   KEY OFF
150  DF$="treedata"                                        'default input file
160  LPM=5                                                  '# of preset trees
170  LP=0                                              'index of current tree
180  P$="Press space bar to "                               'used in prompts
200 REM * initial prompt *
210  MS$="(P)reset data, (D)isk file or (Q)uit: ":                  'prompt
220   GOSUB 6100
230  OP=(1+INSTR("PpDdQq"+CHR$(13),R$))\2               'encode the response
240  CLS
250  ON OP GOTO 270,310,9000,270                                   'branch
270  GOSUB 7000                                            'read preset data
280  GOTO 500                                   'proceed to display the tree
300 REM * get data from disk file
310  MS$="Name of file (default is " + DF$ +"): "                  'prompt
320   GOSUB 6200
330  IF R$<>"" THEN NF$=R$ ELSE NF$=DF$: PRINT NF$;       'specify the file
340  OPEN NF$ FOR INPUT AS #1                                     'open it
350   INPUT #1, IMAX                                           '# of nodes
360  DIM T$(IMAX),P(IMAX),C(IMAX),X(IMAX),Y(IMAX),XIN(IMAX)   'arrays for data
370  CLS
380  LOCATE 10,20
390   PRINT "Please wait - data are being loaded"
400  FOR I=1 TO IMAX                                     'loop through nodes
410   INPUT #1,T$(I)                                            'node name
420   INPUT #1,P(I),C(I),X(I),Y(I)            'parent, child #, coordinates
430  NEXT I
440  CLOSE 1
450  CLS
500 REM * display the tree *
510  DIM D(IMAX),B(IMAX,2),XB(IMAX),YB(IMAX),G(IMAX),IQ(IMAX)
520  FOR I=1 TO IMAX                                     'loop through nodes
530   DRAW "bm =x(i);,=y(i); bu2 r2 d4 l4 u4 r2"              'draw the node
540   V=1+Y(I)\8: H=2+X(I)\8                             'display the label
550    S$=T$(I):  GOSUB 6500
560   IF P(I)=0 THEN 580                                  'jump if root node
570    LINE (X(I),Y(I))-(X(P(I)),Y(P(I)))              'draw edge to parent
580  NEXT I
600 REM * construct descendant counts *
610  FOR I=IMAX TO 1 STEP -1                            'loop through nodes
620   D(P(I))=D(P(I))+D(I)+1            'increase descendant count of parent
630  NEXT I
700 REM * preserve x coordinates to permit repetition *
710  FOR I=1 TO IMAX
720   XIN(I)=X(I)
730  NEXT I
1000 REM * rectify the tree *
1010  FOR I=1 TO IMAX                                   'loop through nodes
1020   MS$=P$ + "deal with edges from node " + T$(I)              'prompt
1030    GOSUB 6000
1040   NC=C(I)                             '# of children of current node
1050    IF NC=0 THEN 2470                                    'jump if none
1060   B(I,1)=I+1                                        'link to left child
1070   XP=X(I)                             'coordinates of current node
1080   YP=Y(I)
```

```
1090    DIM XC(NC),YC(NC),TC$(NC)                        'arrays for its children
1100    IC=I+1                                           'pointer to eldest
1110    FOR JC=1 TO NC                                   'loop through children
1120      XC(JC)=X(IC)                                      'coordinates
1130      YC(JC)=Y(IC)
1140      TC$(JC)=T$(IC)                                       'label
1150      IF JC=NC THEN 1180                               'jump if youngest
1160      B(IC,2)=IC+D(IC)+1             'pointer to right subnode in binary tree
1170      IC=B(IC,2)                                 'pointer to next child
1180    NEXT JC
1500 REM * rectify children except eldest *
1510    FOR JC=NC TO 2 STEP -1                                       'loop
1520      MS$=P$+"break edge from node "+TC$(JC)+" to node "+T$(I)      'prompt
1530      GOSUB 6000
1540      LINE (XP,YP)-(XC(JC),YC(JC)),0,,&HAAAA           'de-emphasize old edge
1550      V=1+YP\8: H=2+XP\8                               'repair the label
1560       S$=T$(I): GOSUB 6500
1570      MS$=P$+"draw edge from node "+TC$(JC)+" to node "+TC$(JC-1)     'prompt
1580      GOSUB 6000
1590      LINE (XC(JC),YC(JC))-(XC(JC-1),YC(JC-1))                 'draw new edge
1600      V=1+YC(JC)\8: H=2+XC(JC)\8                        'repair the label
1610      S$=TC$(JC): GOSUB 6500
1620      MS$=P$+"erase former edge from node "+TC$(JC)+" to node "+T$(I) 'prompt
1630      GOSUB 6000
1640      LINE (XP,YP)-(XC(JC),YC(JC)),0                   'erase the old edge
1650      LINE (XC(JC)+2,YC(JC))-STEP(-8,0)                'repair the new edge
1660      V=1+YP\8: H=2+XP\8                               'repair the labels
1670       S$=T$(I): GOSUB 6500
1680      V=1+YC(JC)\8: H=2+XC(JC)\8
1690      S$=TC$(JC): GOSUB 6500
1700      DRAW "c1 bm =xp;,=yp; bu2 r2 d4 l4 u4 r2"        'repair the nodes
1710      DRAW "c1 bm =xc(jc);,=yc(jc); bu2 r2 d4 l4 u4 r2"
1720    NEXT JC
1730    XC=XC(1)                                     'coordinates of eldest child
1740    YC=YC(1)
1750    ERASE XC,YC,TC$                              'erase arrays for children
2000 REM * align parent of eldest child *
2010    IC=I+1                                               'node # of child
2020    IP=I                                                 'of parent
2030    IG=P(IP)                                             'of grandparent
2040    IF IG=0 THEN 2070
2050    XG=X(IG)                                  'coordinates of grandparent
2060    YG=Y(IG)
2070    IF IP=1 THEN VAC=1                               'parent is root node
2080    IF IP>1 AND IP=IG+1 THEN VAC=2                   'parent is eldest child
2090    IF IG+D(IG)>IP+D(IP) AND IP>IG+1 THEN VAC=3      'parent is a middle child
2100    IF IG+D(IG)=IP+D(IP) AND IP>IG+1 THEN VAC=4      'parent is youngest child
2110    MS$=P$+"break edge from node "
           +T$(IC)+" to node "+T$(IP)                                'prompt
2120    GOSUB 6000
2130    LINE (XC,YC)-(XP,YP),0,,&HAAAA                   'break edge to parent
2140    MS$=P$+"align node "+T$(IP)+" with node "+T$(IC)               'prompt
2150    GOSUB 6000
2160    V=1+YP\8: H=2+XP\8                               'erase label of parent
2170    S$=SPACE$(LEN(T$(IP))): GOSUB 6500
2180    DRAW "c0 bm =xp;,=yp; bu2 r2 d4 l4 u4 r2"        'erase the node
2190    IF VAC>1 THEN
           LINE (XP,YP)-STEP(8*LEN(T$(IP))+8,0)        'repair edge towards right
2200    IF VAC=4 THEN LINE -(XC,YP),0          'erase projection if youngest
2210    IF VAC<>2 THEN 2260                    'jump unless parent is eldest
```

```
222Ø      LINE (XC,YC)-(XC,YP)                'draw vertical edge from child to parent
223Ø      LINE -(XP,YP)                       'draw extra piece of horizontal edge
224Ø      LINE (XP,YP-1)-(XP,YG),Ø,,&HAAAA    'break old parent to grandparent edge
225Ø      LINE -(XC,YP)                       'draw new parent to grandparent edge
226Ø      V=1+YP\8: H=2+XC\8                  'display parent label in new position
227Ø      S$=T$(IP): GOSUB 65ØØ
228Ø      DRAW "c1 bm =xc;,=yp; bu2 r2 d4 14 u4 r2"     'draw node in new position
229Ø      IF VAC<>2 THEN 232Ø                     'jump unless parent is eldest
23ØØ      MS$=P$+"erase old edges"                         'prompt
231Ø        GOSUB 6ØØØ
232Ø      LINE (XP,YP)-(XC,YC),Ø            'erase old edge from parent to child
233Ø      LINE -(XC,YP)                     'draw new edge from child to parent
234Ø        LINE (XC,YC)-STEP(3,Ø)               'repair on left of parent label
235Ø        IF VAC=2 OR VAC=3 THEN LINE (XP-8,YP)-STEP(16,Ø)      'repair on right
236Ø        V=1+YC\8: H=2+XC\8                      'repair label of child
237Ø          S$=T$(IC): GOSUB 65ØØ
238Ø      X(IP)=X(IC)                                   'reset parent coordinate
239Ø      IF VAC<>2 THEN 247Ø                    'jump unless parent is eldest child
24ØØ        LINE (XP,YP-1)-(XP,YG),Ø              'erase old edge to grandparent
241Ø      IC=IP                            'parent becomes child for next cycle
242Ø      IP=IG                                 'grandparent becomes parent
243Ø      YC=YP                                'reset coordinates correspondingly
244Ø      XP=XG
245Ø      YP=YG
246Ø      GOTO 2Ø3Ø                          'loop back to align previous generation
247Ø    NEXT I
3ØØØ REM * display the binary tree
3Ø1Ø    YDIV=Ø                            'initialize value to be set to maximum
3Ø2Ø    FOR I=1 TO IMAX                               'loop through nodes
3Ø3Ø      IF Y(I)>YDIV THEN YDIV=Y(I)                 'reset if necessary
3Ø4Ø    NEXT I
3Ø5Ø    YDIV=8*(YDIV\8)+12                                  'adjust
3Ø6Ø    LINE (Ø,YDIV)-STEP(64Ø,Ø)                 'draw dividing line
3Ø7Ø    MS$=P$+"display the sloping tree"                     'prompt
3Ø8Ø    GOSUB 6ØØØ
3Ø9Ø    LOCATE 25,1
31ØØ    PRINT "Please wait - sorting coordinates" TAB(79)
32ØØ REM * construct generation values *
321Ø    G(1)=1
322Ø    GMAX=1
323Ø    FOR I=1 TO IMAX                               'loop through nodes
324Ø      FOR M=1 TO 2                            'loop through children
325Ø      IF B(I,M)=Ø THEN 329Ø
326Ø      K=B(I,M)
327Ø      G(K)=G(I)+1                  'increase child         's generation #
328Ø      IF G(K)>GMAX THEN GMAX=G(K)
329Ø      NEXT M
33ØØ    NEXT I
331Ø    DIM L(GMAX)                                  'array for edge lengths
34ØØ REM * sort the pointers by generation *
341Ø    FOR Q=1 TO IMAX
342Ø      IQ(Q)=Q                                        'initialize
343Ø    NEXT Q
344Ø    FOR I=IMAX TO 2 STEP -1                            'bubble sort
345Ø      FOR J=1 TO I-1
346Ø      IF G(IQ(J))>G(IQ(J+1)) THEN SWAP IQ(J),IQ(J+1)
347Ø      NEXT J
348Ø    NEXT I
349Ø    FOR G=1 TO GMAX                           'initialize the edge lengths
35ØØ    L(G)=32
```

```
3510    NEXT G
4000 REM * display the binary tree *
4010    XB(1)=320                                           'coordinates of root node
4020    YB(1)=YDIV+8
4030    MS$=P$+"display root node"                                          'prompt
4040     GOSUB 6000
4050    DRAW "c1 bm =xb(1);,=yb(1); bu2 r2 d4 l4 u4 r2"          'draw the node
4060    V=1+YB(1)\8: H=2+XB(1)\8                                        'label it
4070    S$=T$(1): GOSUB 6500
4080    IL=0                                      'direct pointer to predecessor
4090    FOR Q=1 TO IMAX                           'loop through indirect pointers
4100    I=IQ(Q)                                              'direct pointer
4110    IF (YB(I)+L(G(I))/2)\8<23
                OR G(I)=GMAX THEN 4220                       'test for need to scroll
4120    MS$=P$+"scroll the display"                                     'prompt
4130     GOSUB 6000
4140    FOR KS=1 TO L(G(I))/16                    'loop to allow for edge height
4150    LOCATE 25,1                               'scroll one text line
4160     PRINT SPACES$(79)
4170    NEXT KS
4180    FOR J=1 TO IMAX                                      'loop through nodes
4190    YB(J)=YB(J)-L(G(I))/2                                'adjust y coordinates
4200    NEXT J
4210    YDIV=YDIV-L(G(I))/2                       'adjust position of dividing line
4220    IF B(IL,2)>0 AND B(I,1)>0 AND G(I)=G(IL)
                AND L(G(I))=L(G(I)-1) THEN 4500             'test for node overlap
4230    BN=B(I,1)                                           'pointer to left child
4240    IF BN=0 THEN 4330                                    'jump if none
4250    XB(BN)=XB(I)-L(G(I))                                 'form its coordinates
4260    YB(BN)=YB(I)+L(G(I))/2
4270    MS$=P$+"draw edge to node "+T$(BN)                              'prompt
4280     GOSUB 6000
4290    LINE (XB(I),YB(I))-(XB(BN),YB(BN))                   'draw the edge
4300    DRAW "bm =xb(bn);,=yb(bn); bu2 r2 d4 l4 u4 r2"       'draw the node
4310    V=1+YB(BN)\8: H=2+XB(BN)\8                           'label it
4320    S$=T$(BN): GOSUB 6500
4330    BN=B(I,2)                                           'pointer to right child
4340    IF BN=0 THEN 4430                                    'jump if none
4350    XB(BN)=XB(I)+L(G(I))                                 'form its coordinates
4360    YB(BN)=YB(I)+L(G(I))/2
4370    MS$=P$+"draw edge to node "+T$(BN)                              'prompt
4380     GOSUB 6000
4390    LINE (XB(I),YB(I))-(XB(BN),YB(BN))                   'draw the edge
4400    DRAW "bm =xb(bn);,=yb(bn); bu2 r2 d4 l4 u4 r2"       'draw the node
4410    V=1+YB(BN)\8: H=2+XB(BN)\8                           'label it
4420    S$=T$(BN): GOSUB 6500
4430    IL=I                                      'reset pointer to predecessor
4440    NEXT Q
4450    GOTO 5010
4500 REM * increase edge lengths of earlier generations *
4510    FOR G=1 TO G(I)-1                         'loop through previous generations
4520    L(G)=L(G)*3/2                             'increase the edge length
4530    NEXT G
4540    MS$=P$+"redraw tree - nodes will coincide otherwise"            'prompt
4550     GOSUB 6000
4560    VIEW (0,YDIV+1)-(639,199),0               'clear binary tree area
4570    VIEW                                      'reset viewport to entire screen
4580    GOTO 4010                                 'go back to redraw the tree
5000 REM * prompt for further action *
5010    MS$="(D)isplay links, (S)ave, (R)epeat or (M)ain menu: "        'prompt
```

```
5Ø2Ø    GOSUB 61ØØ
5Ø3Ø    ON (1+INSTR("DdSsRrMm",R$))\2 GOTO 51ØØ,54ØØ,56ØØ,56ØØ         'branch
5Ø4Ø    GOTO 5ØØØ                                                     'retry
51ØØ REM * show the connectivity *
511Ø    HT=8*IMAX+2Ø                                           'height of table
512Ø    IF HT>199 THEN HT=199                            'limit to screen depth
513Ø    VIEW (Ø,Ø)-(144,HT),Ø,1                'clear and outline the table are
514Ø    VIEW
515Ø    LOCATE 1,1                                                    'headings
516Ø    PRINT "parent left right "
517Ø    PRINT
518Ø    LINE (Ø,12)-STEP(144,Ø)                             'line below headings
519Ø    LINE (52,Ø)-STEP(Ø,HT)                          'lines between columns
52ØØ    LINE (92,Ø)-STEP(Ø,HT)
521Ø    FOR I=1 TO IMAX                                     'loop through nodes
522Ø    PRINT TAB(2) T$(I);                                              'label
523Ø    LOCATE ,9
524Ø    IF B(I,1)>Ø THEN PRINT T$(B(I,1)); ELSE PRINT "/";          'left child
525Ø    LOCATE ,14
526Ø    IF B(I,2)>Ø THEN PRINT T$(B(I,2)); ELSE PRINT "/";         'right child
527Ø    IF CSRLIN<24 THEN 533Ø                        'repair lines if scrolled
528Ø    LINE (52,Ø)-STEP(Ø,HT)                          'lines between columns
529Ø    LINE (92,Ø)-STEP(Ø,HT)
53ØØ    LINE (144,Ø)-STEP(Ø,HT)
531Ø    LINE (145,Ø)-STEP(Ø,HT)
532Ø    LINE -STEP(-145,Ø)
533Ø    NEXT I
534Ø    GOTO 5ØØØ
54ØØ REM * save the binary tree on disk *
541Ø    MS$="Name of output file: "                                    'prompt
542Ø    GOSUB 62ØØ
543Ø    NB$=R$
544Ø    OPEN NB$ FOR OUTPUT AS #2                                'open the file
545Ø    PRINT #2, IMAX                                            '# of nodes
546Ø    FOR I=1 TO IMAX                                     'loop through nodes
547Ø    PRINT #2, T$(I)                                                 'label
548Ø    PRINT #2, B(I,1),B(I,2),XB(I),YB(I)-YB(1)   'children and coordinates
549Ø    NEXT I
55ØØ    CLOSE 2
551Ø    GOTO 5Ø1Ø
56ØØ REM * prepare to repeat or return to main menu *
561Ø    CLS                                                 'clear the screen
562Ø    ERASE D,B,XB,YB,G,IQ,L                    'erase arrays for binary tree
563Ø    IF INSTR("Mm",R$) THEN 568Ø
564Ø    FOR I=1 TO IMAX                       'reset x coords to original values
565Ø    X(I)=XIN(I)
566Ø    NEXT I
567Ø    GOTO 5ØØ
568Ø    ERASE T$,P,C,X,Y,XIN                     'erase arrays for input tree
569Ø    GOTO 2ØØ                                    'go back to process another
6ØØØ REM * prompt for a key press to single cycle *
6Ø1Ø    GOSUB 63ØØ                                        'display the message
6Ø2Ø    Z$=INPUT$(1)                                        'wait for key press
6Ø3Ø    RETURN
61ØØ REM * prompt for one-character response *
611Ø    GOSUB 63ØØ                                        'display the message
612Ø    R$=INPUT$(1)                                        'store a key press
613Ø    IF R$<>CHR$(13) THEN PRINT R$;                 'display unless Enter
614Ø    RETURN
62ØØ REM * prompt for a string *
```

```
6210    GOSUB 6300                                                  'display the message
6220    INPUT; "", R$                                               'store the response
6230    RETURN
6300 REM * display a prompt *
6310    LOCATE 25,1                                                 'clear the prompt line
6320    PRINT SPC(79)
6330    LOCATE 25,1                                                 'display the message
6340    PRINT MS$;
6350    RETURN
6500 REM * display a string if in range *
6510    IF 0>V OR V>24 OR 0>H OR H+LEN(S$)>80 THEN 6540             'check in range
6520    LOCATE V,H                                                  'display it
6530    PRINT S$
6540    RETURN
7000 REM * set up demonstration data *
7010    IF LP=LPM THEN RESTORE                                      'restore if final set was used
7020    LP=1+(LP MOD LPM)                                           'increase or reset the set #
7030    READ IMAX                                                   '# of nodes
7040    DIM T$(IMAX),P(IMAX),C(IMAX),X(IMAX),Y(IMAX),XIN(IMAX)      'arrays for data
7050    FOR I=1 TO IMAX                                             'loop through nodes
7060       READ T$(I),P(I),C(I),X(I),Y(I)      'label, parent, # of children & coords
7070    NEXT I
7080    RETURN
7100 REM * preset data *
7110    DATA 5,a,0,4,320,12,b,1,0,80,98,c,1,0,240,98,d,1,0,400,98,e,1,0,560,98
7120    DATA 9,a,0,2,480,10,b,1,2,380,30,d,2,2,280,50,f,3,2,180,70
7130    DATA   h,4,0,80,90,i,4,0,280,90,g,3,0,380,70,e,2,0,480,50,c,1,0,580,30
7140    DATA 7,a,0,2,320,10,b,1,2,160,30,d,2,0,80,50
7150    DATA   e,2,0,240,50,c,1,2,480,30,f,5,0,400,50,g,5,0,560,50
7160    DATA 10,a,0,3,320,12,b,1,2,160,60,e,2,0,120,84,f,2,0,200,84
7170    DATA   c,1,2,320,60,g,5,0,280,84,h,5,0,360,84,d,1,2,480,60
7180    DATA   i,8,0,440,84,j,8,0,520,84
7190    DATA 29,a,0,2,40,4, b,1,0,8,12, c,1,2,72,12, d,3,0,40,20, e,3,2,104,20
7200    DATA   f,5,0,72,28, g,5,2,136,28, h,7,0,104,36, i,7,2,168,36
7210    DATA   j,9,0,136,44, k,9,2,200,44, l,11,0,168,52, m,11,2,232,52
7220    DATA   n,13,0,200,60, o,13,2,264,60, p,15,0,232,68, q,15,2,296,68
7230    DATA   r,17,0,264,76, s,17,2,328,76, t,19,0,296,84, u,19,2,360,84
7240    DATA   v,21,0,328,92, w,21,2,392,92, x,23,0,360,100, y,23,2,424,100
7250    DATA   z,25,0,392,108, A,25,2,456,108, B,27,0,424,116, C,27,0,488,116
8000 REM * error trap *
8010    MS$="Error"+STR$(ERR)+" in line"+STR$(ERL)+" - press space bar"
8020       GOSUB 6000                                               'diagnostic
8030    RESUME 9000
9000 REM * quit *
9010    SCREEN 0                                                    'reset the screen
9020    CLS
9030    CLOSE 1,2                                                   'close files
9040    ON ERROR GOTO 0                                             'reset error address
9050    END                                                         'quit
```

L(G)   The length of the horizontal projection of the edges from the gth generation nodes to their children

R$ The one-byte response to the prompts for certain options

P$ The string "Press space bar to" that occurs at the beginning of several prompts

MS$ The text of the current prompt

KS
:   The index of the loop that scrolls the picture when it has reached the bottom of the screen by the number of text lines appropriate to the length of the edges that will be drawn next

NB$
:   The name of the output file of data that represents the binary tree

HT
:   The height of the table of links that is displayed when D is typed in response to the prompt for further action (HT corresponds to the number of nodes unless there are more than 22, in which case it is just the height of the screen)

V,H,S$
:   The vertical and horizontal cursor positions and the string passed to the subroutine that displays the names of nodes

XIN(i)
:   The initial value of the x coordinate of the ith node preserved for use by the R option

Program Notes

The internal operation of BINCON follows the description of its external behavior in a direct fashion. After a tree has been rectified, the array of generation values G(i) is constructed. Then IQ, the array of pointers to the nodes of the binary tree, arranged by generation, is formed by a simple sort. The binary tree is drawn by a FOR ... NEXT loop that works through the nodes in the order specified by the successive entries in the IQ array, that is, from left to right in successive generations.

```
4090 FOR Q=1 TO IMAX
4100   I=IQ(Q)
. . .
4440 NEXT Q
```

Node I is drawn and labeled. The program scrolls and/or increases the edge lengths if necessary. In general, it proceeds directly to set BN to the pointer to the left child. Unless this is zero, the coordinates of the child are computed and the appropriate edge and node are drawn and labeled. The program then uses the pointer to the right child in the corresponding manner.

Suggested Activities

1. Expand BINCON to show on the screen how the entries in the tabular representation of the general tree are used by the al-

gorithmic basis of the program and how the entries in the tabular representation of the binary tree are formed and used.

2. Write a program that (a) reads the representation of a binary tree produced by BINCON, (b) displays the binary tree, and (c) constructs and displays the general tree that it represents.

3. BINCON rectifies a tree by dealing with the children of a node from right to left, but it works through the parents in top-to-bottom, left-to-right order. This follows naturally from the representation that we have used for a general tree. Expand BINCON to work through the nodes in each generation, treated as parents, in right-to-left order.

5.4   Traversing a Binary Tree

Throughout the last three sections the systematic inspection of the nodes of a tree has progressed in top-to-bottom, left-to-right order. This is also called *preorder traversal* of a tree, and it can be defined recursively as follows:

1. Visit the root node.

2. Apply preorder traversal to the subtrees of the root node.

3. Deal with the subtrees in order from left to right.

For a binary tree, the definition of preorder traversal becomes:

1. Visit the root node.

2. Apply preorder traversal to the left subtree of the root node, if it has one.

3. Apply preorder traversal to the right subtree of the root node, if it has one.

Preorder traversal is very convenient for many applications of trees, but some other orders of traversal are preferred for particular purposes. The program BINTRAVL demonstrates the three commonest traversals. These are preorder traversal and the inorder and postorder traversals that are explained shortly.

The program begins by prompting you to type P for preset data, D for data in a disk file, or Q to quit. If you type P for preset data, the screen clears and the following binary tree is displayed with its root near the top center of the screen.

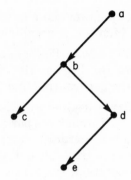

The program then prompts you to type 1 for preorder, 2 for inorder, and 3 for postorder traversal, respectively. If you type 1, for preorder traversal, the program displays the words "preorder traversal:" at the lower-left corner of the screen. The words "name" and "side" that head the two columns of an "upside down" stack appear at the top-right corner of the screen. The program enlarges node a of the tree and prompts you to include node a in the output. The node name a is put into the record of the traversal at the bottom of the screen. The name is also put into the stack.

```
name    side
  a
```

Then you are prompted to test if node a has a left child. You respond. The indicator L appears alongside the node name a in the stack to show that a left subtree is being explored. The edge from node a to node b is intensified. Node b is enlarged, and you are prompted to include node b in the output. You respond. The node name b is appended to the record of the traversal that is displayed at the bottom of the screen. The name is pushed onto the stack.

```
name    side
  a      L
  b
```

The sequence of events that led from the inclusion of node a to the inclusion of node b now is paralleled to lead to the inclusion of node c. The stack becomes:

```
name    side
  a      L
  b      L
  c
```

The record of the traversal becomes:

```
Preorder traversal: a b c
```

Then you are prompted in turn to test if node c has a left child, to test if node c has a right child, and to backtrack from node c. The node c and the edge from node b to node c are deemphasized. The row of the stack containing the name c is erased.

```
name    side
  a     L
  b     L
```

You are prompted to test if node b has a right child. You respond, and the edge from node b to node d is intensified. Node d is enlarged and the name d is appended to the output line. The indicator for node b changes to R, and the name d is pushed onto the stack.

```
name    side
  a     L
  b     R
  d
```

A sequence of prompts similar to those already described leads to the intensification of the edge de and the inclusion of the indicator L alongside d in the stack. The name e is pushed onto the stack:

```
name    side
  a     L
  b     R
  d     L
  e
```

Node e is appended to the record of the traversal:

```
Preorder traversal: a b c d e
```

Further prompts that are similar to those already described lead to (1) the deemphasis of node e, (2) removal of node e from the stack, (3) the deemphasis of edge de, (4) the deemphasis of node d, (5) removal of node d from the stack, (6) the deemphasis of edge bd, (7) the deemphasis of node b, (8) removal of node b from the stack, and (9) the deemphasis of edge ab.

The program then prompts you to test if node a has a right child. This test is redundant when the binary tree represents a general tree, but binary trees in which the root has two children do occur. For the examples in hand, however, there is no right child. The next prompt tells you the traversal has ended. The name a is erased from the stack, and the prompt changes to type T to traverse the same tree again, N to traverse a new tree, or Q to quit.

The program provides several sets of preset data that display different trees for traversal. These are brought into use on the successive occasions when preset data is requested. The sequence then repeats when further requests are made. If you type D, the program prompts for the name of the input file. You enter the name of a file that has been created by BINCON. The program loads the file, displays the tree, and prompts for the type of traversal.

Turning now to the inorder and postorder traversals, the recursive definition of inorder traversal of a binary tree is as follows.

1. Apply inorder traversal to the left subtree of the root node, if it has one.

2. Visit the root node.

3. Apply inorder traversal to the right subtree of the root node, if it has one.

The definition of postorder traversal is:

1. Apply postorder traversal to the left subtree of the root node, if it has one.

2. Apply postorder traversal to the right subtree of the root node, if it has one.

3. Visit the root node.

Preorder traversal thus visits the root node before traversing the subtrees. Inorder traversal visits the root node between subtree traversals. Postorder traversal visits the root node after traversing the subtrees.

The BINTRAVL program lets you trace inorder and postorder traversals in a manner completely analogous to that described for preorder traversal. The same prompts appear, but in a different order. Nodes and edges are emphasized when they come under attention and deemphasized when they have been included in the traversal. For the tree used to illustrate preorder traversal, that is,

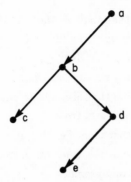

the inorder traversal is c b e d a and the postorder traversal is c e d b a. The preset data in BINTRAVL also defines the following three trees.

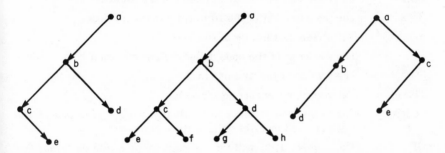

The traversals of these trees are respectively as follows:

Preorder	abced	abcefdgh	abdce
Inorder	cebda	ecfbgdha	dbaec
Postorder	ecdba	efcghdba	dbeca

In the binary tree representation of a general tree, the root node has only a left child. You can represent a "forest," that is, a collection of trees, by arranging them in order and making the root node of each original tree the right child of the root of the preceding tree. Binary trees also occur in many ways besides representing general trees and forests, for example, when you deal with algebraic expressions (see Suggested Activities).

Program Variables

MMAX   The number of prompting messages

M$(m,l)   The lth line of the mth prompting message, stored in a compact form

NF$	The name of the input file that specifies a tree
DF$	The default value of the file name
DS1$	The draw string that displays a node initially
DS2$	The draw string that redisplays a node in its initial size
DS3$	The draw string that displays an enlarged node
LPM	The number of preset trees
LP	The number that identifies the preset tree in current use
M	The number that identifies the current prompt
R$	The response to a prompt
IMAX	The number of nodes in the current tree
T$(i)	The name of the ith node
B(i,1)	The node number of the left child of the ith node
B(i,2)	The node number of the right child of the ith node
X(i),Y(i)	The screen coordinates of the ith node
I	The subscript of the node under current attention
S(j)	The I value of the jth node in the stack
TR$	The record of the current traversal
LOR$(j)	The letter L or R that shows whether the $(j+1)$th node in the stack is a left or right child of its parent
OP	The number 1, 2, or 3 for preorder, inorder, and postorder traversal, respectively
J	The pointer to the top of the stack
BN	The I value of the child of the node at the top of the stack under current attention
LMM(m)	The number of lines in the mth prompt
LM	The number that identifies the current line
QS	The position of an asterisk, in the compact form of a prompt line, that is replaced by the name of a node in the actual display
CMSL$	The line that is actually displayed

Program Notes

The algorithmic basis of the traversal is handled very simply. Thus, for preorder traversal, the stack pointer J is initialized to 1, the pointer to the root node is pushed onto the stack, and the record of the traversal is initialized to the name of the root node. The repetitive part of the algorithm is started by prompting for a key press and setting BN to the I value of the left child of the node at the top of the stack.

```
1200   M=4:  GOSUB 6000
1210   BN=B(S(J),1)
```

If it is greater than 0, the left-or-right pointer is set to L and associated with the current node on the top of the stack and the pointer BN is pushed onto the next row of the stack.

```
1220   IF BN=0 THEN 1400
1230   GOSUB 8000
. . .
8010   LOR$(J)="L"
8020   GOSUB 8500
. . .
8540   J=J+1
8550   S(J)=BN
```

The child is appended to the record of the traversal, and the cycle is repeated.

```
1310   TR$=TR$+" "+T$(S(J))
1320   GOSUB 7500
1330   GOTO 1200
```

If the node under consideration at the start of the cycle does not have a left child, BN is set to the I value of its right child. If the I value is greater than zero, the left-or-right indicator is set to R and the pointer to the child is pushed onto the stack.

```
1410   BN=B(S(J),2)
1420   IF BN=0 THEN 1500
1430   GOSUB 8200
. . .
8210   LOR$(J)="R"
8220   GOSUB 8500
```

The child is appended to the traversal and the cycle is repeated. If the node under consideration at the start of the cycle has no children, the stack is popped.

```
1500   GOSUB 8800
. . .
8910   J=J-1
. . .
8930   IF J=0 THEN RETURN 4000 ELSE RETURN
```

If this does not empty the stack, the program branches on the left-or-right indicator of the node that is now at the top. If that is L, the program goes back to test if there is a right child, and to deal with this in the manner that has been described. If, however, the indicator of the exposed node is R, the program pops the stack again.

The algorithmic basis of the inorder traversal first pushes the root node onto the stack. The repetitive part then tests if there is a left child and, if so, sets the left-or-right indicator to L, pushes the pointer to the left child onto the stack, and starts the cycle afresh (lines 2100 to 2140). If the node under attention does not have a left child, it is appended to the record of the traversal. The program then tests if there is a right child and, if there is, sets the left-or-right indicator to R, pushes the pointer to the child onto the stack, and starts the cycle again (lines 2310 to 2340).

If there is no right child, the stack is popped. If that does not empty the stack, the program branches on the left-or-right indicator of the node that is now at the top of the stack. If that is L, the program appends the node to the record of the traversal and tests for a right child. If the indicator is R, the program pops the stack again (lines 2400 to 2410).

The algorithmic basis of the postorder traversal begins by pushing the root node onto the stack. The repetitive part then tests if there is a left child and, if there is one, sets the left-or-right indicator to L, pushes the pointer to the child onto the stack, and starts the cycle afresh (lines 3100 to 3140). If there is no left child, the corresponding actions are taken for the right child (lines 3200 to 3240). If the node under current attention does not have a right child, its name is appended to the record of the traversal. This is redisplayed and the stack is popped (lines 3310 to 3400). If this does not empty the stack, the program branches on the left-or-right indicator of the node that is at the top. If that is L, the program tests whether the node has a right child. If a right child exists, it is pushed onto the stack and the cycle is repeated. Otherwise, the program appends the name of the node to the traversal and pops the stack again (line 3410).

Suggested Activities

1. A reverse preorder traversal of a binary tree (1) visits the root node, (2) traverses the right subtree in reverse preorder, (3) traverses the left subtree in reverse preorder. The reverse inorder and postorder traversals are defined in the corresponding fashions by switching the words left and right in the definitions of the ordinary inorder and postorder traversals. Extend the BINTRAVL program to perform these reverse preorder, inorder, and postorder traversals.

2. Write a program that lets you specify a binary tree as follows. First, you enter the name of the root node. In each further cycle of the

```
100 REM * BINTRAVL traverses a binary tree *
110   ON ERROR GOTO 9000                                      'set error trap
120   SCREEN 2                                                'screen set up
130    KEY OFF
140    CLS
150   LOCATE 10,20
160    PRINT "Please wait - data loading"
170   MMAX=8                                                   '# of prompts
180   DIM MS$(MMAX,3)                                     'array of messages
190   LPM=4                                              '# of preset trees
200   LP=0                                           'index of current tree
210   DF$="bindata"                                     'default input file
220   DS1$="bm=x(i);,=y(i); bu2 r2 d4 14 u4 r2 bd2"   'draws node initially
230   DS2$="bm =x(s(j));, =y(s(j)); bu2 r2 d4 14 u4 r2 bd2"       'redraws it
240   DS3$="bm =x(s(j));, =y(s(j)); bu2 r2 d4 18 u4 r6 bd2"      'enlarges it
250   ORD$(1)="Pre": ORD$(2)="In": ORD$(3)="Post"
260   GOSUB 5000
300 REM * prompt for action *
310   CLS                                              'clear the screen
320   M=1: GOSUB 6000                                            'prompt
330    CLS                                             'clear the screen
340   ON (1+INSTR("PpDdQq",R$))\2 GOTO 400,600,10000   'branch on response
400 REM * read preset data *
410   IF LP=LPM THEN RESTORE 5310            'restore if final set used
420   LP=1+(LP MOD LPM)                      'increase or reset counter
430   READ IMAX                                                '# of nodes
440   DIM T$(IMAX),B(IMAX,2),X(IMAX),Y(IMAX)           'arrays for data
450   FOR I=1 TO IMAX                                   'loop through nodes
460     READ T$(I),B(I,1),B(I,2),X(I),Y(I)      'label, children & coords
470     Y(I)=Y(I)+12                             'increase y coordinate
480   NEXT I
490   GOSUB 7000                                              'draw the tree
500   GOTO 800
600 REM * read a disk file *
610   PRINT "Name of input file (default is " DF$ "): ";          'prompt
620   INPUT; "", NF$
630   IF NF$<>"" THEN PRINT ELSE NF$=DF$: PRINT NF$        'file name
640   CLS                                              'clear screen
650   OPEN NF$ FOR INPUT AS #1                           'open file
660   INPUT #1, IMAX                                      '# of nodes
670   DIM T$(IMAX),B(IMAX,2),X(IMAX),Y(IMAX)           'arrays for data
680   FOR I=1 TO IMAX                                   'loop through nodes
690     INPUT #1, T$(I)                                        'label
700     INPUT #1, B(I,1),B(I,2),X(I),Y(I)        'children & coordinates
710     Y(I)=Y(I)+4                              'adjust y value
720   NEXT I
730   GOSUB 7000                                              'draw the tree
740   CLOSE 1                                           'close the file
800 REM * prepare for traversal *
810   DIM S(IMAX),LOR$(IMAX)                              'stack arrays
820   M=2: GOSUB 6000                                            'prompt
830    OP=VAL(R$)                                      'selected order
840   GOSUB 7500
850   ON OP GOTO 1000,2000,3000                          'branch on it
860   GOTO 820                                                   'retry
1000 REM * preorder traversal *
1010   GOSUB 7200                                     'select the root node
1100   M=3: GOSUB 6000                                          'prompt
1110    TR$=T$(1)                                'include it in traversal
1120    GOSUB 7500
```

```
1200   M=4: GOSUB 6000                                                      'prompt
1210    BN=B(S(J),1)                                              'select left child
1220     IF BN=0 THEN 1400                                           'jump if none
1230      GOSUB 8000            'push L onto stack and emphasize edge & node
1300   M=3: GOSUB 6000                                                      'prompt
1310    TR$=TR$+" "+T$(S(J))                             'include node in traversal
1320     GOSUB 7500
1330     GOTO 1200                              'go back to look for its left child
1400   M=5: GOSUB 6000                                                      'prompt
1410    BN=B(S(J),2)                                             'select right child
1420     IF BN=0 THEN 1500                                           'jump if none
1430      GOSUB 8200           'push R onto stack and emphasize edge & node
1440      GOTO 1300                             'go back to deal with new node
1500   GOSUB 8800                                                     'pop the stack
1510   IF LOR$(J)="L" THEN 1400 ELSE 1500                        'branch on L or R
2000 REM * inorder traversal *
2010   GOSUB 7200                                         'push root node onto stack
2100   M=4: GOSUB 6000                                                      'prompt
2110    BN=B(S(J),1)                                                   'left child
2120     IF BN=0 THEN 2200                                           'jump if none
2130      GOSUB 8000            'push L onto stack and emphasize edge & node
2140      GOTO 2100                             'go back to deal with new node
2200   M=3: GOSUB 6000                                                      'prompt
2210    TR$=TR$+" "+T$(S(J))                             'include node in traversal
2220     GOSUB 7500
2300   M=5: GOSUB 6000                                                      'prompt
2310    BN=B(S(J),2)                                                  'right child
2320     IF BN=0 THEN 2400                                           'jump if none
2330      GOSUB 8200           'push R onto stack and emphasize edge & node
2340      GOTO 2100                             'go back to deal with new node
2400   GOSUB 8800                                                     'pop the stack
2410   IF LOR$(J)="L" THEN 2200 ELSE 2400                        'branch on L or R
3000 REM * postorder traversal *
3010   GOSUB 7200                                             'stack the root node
3100   M=4: GOSUB 6000                                                      'prompt
3110    BN=B(S(J),1)                                                   'left child
3120     IF BN=0 THEN 3200                                           'branch if none
3130      GOSUB 8000            'push L onto stack and emphasize edge & node
3140      GOTO 3100                             'go back to deal with new node
3200   M=5: GOSUB 6000                                                      'prompt
3210    BN=B(S(J),2)                                                  'right child
3220     IF BN=0 THEN 3300                                           'jump if none
3230      GOSUB 8200           'push R onto stack and emphasize edge & node
3240      GOTO 3100                             'go back to deal with new node
3300   M=3: GOSUB 6000                                                      'prompt
3310    TR$=TR$+" "+T$(S(J))                             'include node in traversal
3320     GOSUB 7500
3400   GOSUB 8800                                                     'pop the stack
3410   IF LOR$(J)="L" THEN 3200 ELSE 3300                        'branch on L or R
4000 REM * prompt for further action *
4010   ERASE S,LOR$                                                'erase the stack
4020   TR$=""                        'initialize the record of the traversal
4030   M=8: GOSUB 6000                                                      'prompt
4040   ON (1+INSTR("TtNnQq",R$))\2 GOTO 4050,4080,10000             'branch
4050    TR$=""                                       'nullify record of traversal
4060    GOSUB 7500
4070    GOTO 800                                       'go back to traverse again
4080    ERASE T$,B,X,Y                                  'erase data for tree
4090    GOTO 310                                        'go back to get another
5000 REM * store the prompt messages *
```

```
5010    FOR M=1 TO MMAX                            'loop through the messages
5020      READ LMM(M)                                      '# of lines
5030      FOR LM=1 TO LMM(M)                        'loop through lines
5040        READ MS$(M,LM)                          'transfer the line
5050      NEXT LM
5060    NEXT M
5070    RETURN
5100 REM * text of prompt messages *
5110    DATA 3,"(P)reset data", "(D)isk file.", "or (Q)uit: "
5120    DATA 3,"Type 1 for preorder,", "    2 for inorder,"
5130    DATA   "    3 for postorder: "
5140    DATA 3,!, "include node *", "in the output"
5150    DATA 3,!, "test if node *", "has a left child"
5160    DATA 3,!, "test if node *", "has a right child"
5170    DATA 2,!, "backtrack from node *"
5180    DATA 2,!, "end the traversal"
5190    DATA 3,"(T)raverse again,", "(N)ew tree, or", "(Q)uit: "
5300 REM * preset tree data *
5310    DATA 5,a,2,0,320,0, b,3,4,272,24, c,0,0,224,48, d,5,0,320,48
5320    DATA   e,0,0,272,72
5330    DATA 5,a,2,0,320,0, b,3,4,272,24, c,0,5,224,48, d,0,0,320,48
5340    DATA   e,0,0,272,72
5350    DATA 8,a,2,0,320,0, b,3,4,248,36, c,5,6,176,72, d,7,8,320,72
5360    DATA   e,0,0,128,96, f,0,0,224,96, g,0,0,272,96, h,0,0,368,96
5370    DATA 5,a,2,3,320,0, b,4,0,272,24, c,5,0,368,24, d,0,0,224,48
5380    DATA   e,0,0,320,48
6000 REM * display a prompt *
6010    VIEW (0,0)-(199,23),0                              'clear the message area
6020    VIEW
6030    FOR LM=1 TO LMM(M)                         'loop through lines of message
6040      CMSL$=MS$(M,LM)                                  'tentative line
6050      IF CMSL$<>"!" THEN 6080                  'use standard first line
6060        CMSL$="Press space bar to"                     'when designated by !
6065        SW=1
6070        GOTO 6100
6080      QS=INSTR(CMSL$,"*")                              'search for asterisk
6090      IF QS>0 THEN MID$(CMSL$,QS,1)=T$(S(J))           'replace by node name
6100      LOCATE LM,1
6110      PRINT CMSL$;                                     'display the line
6120    NEXT LM
6125    IF SW THEN SW=0: Z$=INPUT$(1): RETURN      'allow for continuous demo
6130    R$=INPUT$(1)                                       'store key press
6140    IF R$<>CHR$(13) THEN PRINT R$ ELSE PRINT   'display it unless Enter
6150    RETURN
7000 REM * draw the tree *
7010    FOR I=1 TO IMAX                            'loop through nodes
7020      V=1+Y(I)\8: H=2+X(I)\8
7030      IF 0>V OR V>24 OR 0>H OR H+LEN(T$(I))>80 THEN 7100   'check in range
7040      LOCATE V,H                               'display the label
7050      PRINT T$(I)
7060      DRAW "c0" + DS1$ + "p0,0" + "c1" + DS1$ + "p1,1"     'draw the node
7070      IF B(I,1)>0 THEN
            LINE (X(I),Y(I))-(X(B(I,1)),Y(B(I,1))),,,&HAAAA   'left edge
7080      IF B(I,2)>0 THEN
            LINE (X(I),Y(I))-(X(B(I,2)),Y(B(I,2))),,,&HAAAA   'right edge
7090    NEXT I
7100    RETURN
7200 REM * process the root node *
7210    J=1                                        'push it onto stack
7220    S(J)=1
```

```
7230   DRAW "c0" + DS2$ + "p0,0" + "c1" + DS3$ + "p1,1"              'draw it
7240   LOCATE 1,69
7250    PRINT "name side"                               'heading of stack
7260  LOCATE J+1,71
7270    PRINT T$(S(J))                                        'root node
7280   RETURN
7500 REM * display list of nodes in order of traversal *
7510   LOCATE 25,1
7520    PRINT ORD$(OP) "order traversal: " RIGHT$(TR$,67) TAB(79)
'path
7530   RETURN
8000 REM * process a left child *
8010   LOR$(J)="L"                                      'set the indicator
8020   GOSUB 8500                              'common action for children
8030   RETURN
8200 REM * process a right child *
8210   LOR$(J)="R"                                      'set the indicator
8220   GOSUB 8500                              'common action for children
8230   RETURN
8500 REM * push data for a node and emphasize it *
8510   IF J>23 THEN 8540                           'display the indicator
8520   LOCATE J+1,75                                    'if there is room
8530    PRINT LOR$(J);
8540   J=J+1                                 'push node name onto stack
8550   S(J)=BN
8560   IF J>23 THEN 8590                           'display the node name
8570   LOCATE J+1,71                                    'if there is room
8580    PRINT T$(S(J));
8590   LINE (X(S(J-1))-1,Y(S(J-1)))-(X(S(J))-1,Y(S(J))),1    'emphasize the edge
8600   LINE (X(S(J-1)),Y(S(J-1)))-(X(S(J)),Y(S(J))),1
8610   LINE (X(S(J-1))+1,Y(S(J-1)))-(X(S(J))+1,Y(S(J))),1
8620   DRAW "c0" + DS2$ + "p0,0" + "c1" + DS3$ + "p1,1"         'and the node
8630   RETURN
8800 REM * pop the stack *
8810   M=7+(J>1): GOSUB 6000                                         'prompt
8820   IF J=1 THEN 8870
8830    LINE (X(S(J-1))-1,Y(S(J-1)))-(X(S(J))-1,Y(S(J))),0         'erase the
8840    LINE (X(S(J-1)),Y(S(J-1)))-(X(S(J)),Y(S(J))),0            'thickened
8850    LINE (X(S(J-1))+1,Y(S(J-1)))-(X(S(J))+1,Y(S(J))),0             'line
8860    LINE (X(S(J-1)),Y(S(J-1)))-(X(S(J)),Y(S(J))),1,,&HAAAA   'redraw dashed
8870   DRAW "c0" + DS3$ + "p0,0" + "c1" + DS2$ + "p1,1"    'deemphasize child
8880   IF J>23 THEN 8910
8890    LOCATE J+1,71                                    'erase from screen
8900    PRINT SPC(6)
8910   J=J-1                                 'decrease stack pointer
8920   DRAW "c0" + DS3$ + "p0,0" + "c1" + DS3$ + "p1,1"     'repair parent node
8930   IF J=0 THEN RETURN 4000 ELSE RETURN               'branch if empty
9000 REM * error trap *
9010   LOCATE 25,1                                          'diagnostic
9020    PRINT "Error" ERR " in line" ERL " - press space bar";
9030    Z$=INPUT$(1)
9040    RESUME 10000
10000 REM * quit *
10010   CLOSE 1
10020   SCREEN 0
10030   ON ERROR GOTO 0
10040   END
```

input process, you enter the name of a child of a node that has been stored already and specify whether the new node is the left or right child of its parent. The program should prevent you from putting a node into a position that is already occupied. Represent the tree by an array of three columns. The first item in a row of the array is the name of the node that the row represents. The second item points to the row containing the data for the left child of the node, or 0 if the node does not have a left child. The third item deals in a similar manner with the right child. Make the program display the array, and save it on disk when you request these actions in response to a prompt.

3. You can represent an algebraic expression as a binary tree in which the terminal nodes are the numbers and variables and each binary operator is the parent of the two subtrees that represent the expressions which it combines. Use the program described in Activity 2 to store the binary tree of an algebraic expression and produce the preorder, inorder, and postorder traversals. Compare them with the prefix, infix, and postfix forms of the expression.

4. Expand the program described in Activity 2 to display a diagram of the tree. Use the methods of Section 5.3 to keep the diagram compact without letting nodes coincide on the screen.

5. Expand the program to let you delete as well as insert terminal nodes.

6. Expand the deletion process to let you specify the root of a subtree that the program then deletes.

7. Use the linked list principle to keep track of the free space as well as the connections between nodes of the tree and to let the program use this space.

5.5 Finding the Length of the Shortest Path

Suppose that four towns are numbered 1 to 4 and that one-way roads go directly from town 1 to towns 2 and 3, from town 3 to towns 2 and 4, and from town 2 to town 4. Suppose that the distances along these roads are as follows: 30 miles from 1 to 2, 10 miles from 1 to 3, 25 miles from 2 to 4, 15 miles from 3 to 2, and 20 miles from 3 to 4. Thus, the direct road from 1 to 2 winds much more than the roads from 1 to 3 and from 3 to 2. The net effect is shown by the accompanying illustration.

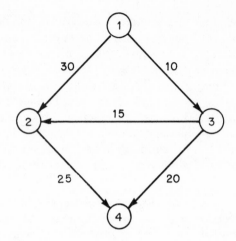

A traveler can get from town 1 to town 4 in several ways. The route via town 2 is 55 miles long. The route via town 3 is 30 miles long. The route via towns 3 and 2 is 50 miles long. Thus the shortest route is 30 miles long.

When the number of towns is very small, the shortest route from one to another can be found by inspection after all the possible routes and their lengths are listed. This approach becomes much too tedious, however, when the number of towns is very large. The same problem occurs when you want to find the minimum travel time or shipping costs between pairs of towns that have a direct or indirect connection. In every case, the problem can be represented by a weighted graph, in the terminology explained at the beginning of this chapter.

Because the graph consists of one-way edges, it is also said to be *directed*. Two nodes cannot be connected by more than one edge going in the same direction. Two nodes can be connected, however, by edges going in opposite directions. These edges can have the same length, but that is not necessary. A path between two nodes is simply a succession of edges that connect them. The length of the path is just the sum of the lengths of the constituent edges. The nodes can be towns and the edge lengths can be mileages, but they need not be. The mathematical problem is the same when the lengths are travel times or transportation costs or any other additive properties of the edges. Whatever the nature of the nodes and the nature of the lengths, the problem is described as finding the shortest path. When we speak about edge lengths from here on, this generality will be taken for granted.

The program FSPATH deals with the shortest path problem in a systematic manner. It lets you specify the relevant details of a weighted graph such as the one shown on page 246. Then it lets you specify a starting node and finds the lengths of the shortest paths from that node to all the other nodes. You press the space bar to make the program proceed from each step to the next. It displays a detailed explanation of the algorithm as it goes along.

FSPATH first clears the screen, draws a vertical line down the center, and prompts you to type D for an automatic demonstration, I to input your own data, or Q to quit. If you type D, the graph shown on page 246 appears above the prompt, on the left side of the screen. You are prompted to enter the number of the node from which you want shortest paths computed. Suppose that you type 1. You are prompted to press the space bar to use the arcs from the starting node 1.

You respond, and the outline of the trial value table appears in the top-right corner of the screen. This table plays a central role in the algorithm.

trial values of shortest paths from 1			
node no.	shortest path yet	length of arc from 1	shortest path now

The contents of this table and, in due course, of another table, change from step to step when you press the space bar in response to successive prompts that are displayed in the lower-left corner of the screen. The original graph stays unchanged on the left side of the screen above the prompt area.

The program works outward from the starting node that you specified, in this case node 1. First, it prompts you to press the space bar to copy the lengths of the arcs from node 1 to node 2 into the trial value table. The program finds that node 1 has a direct connection to node 2 that is 30 miles long. This information is displayed in the table.

trial values of shortest paths from 1			
node no.	shortest path yet	length of arc from 1	shortest path now
2		30	

The prompt changes to "Press space bar to use it as the best length to node 2 so far." The table changes to:

trial values of shortest paths from 1			
node no.	shortest path yet	length of arc from 1	shortest path now
2		30	30

The prompt changes to "Press space bar to copy length of arc from node 1 to node 3 into the trial value table." The program finds that node 1 has a direct connection to node 3; it is 10 units long. This information is displayed in the table, too.

trial values of shortest paths from 1			
node no.	shortest path yet	length of arc from 1	shortest path now
2		30	30
3		10	

The prompt changes, just as it did before. Correspondingly, the 10 is copied into the fourth column of the table:

trial values of shortest paths from 1			
node no.	shortest path yet	length of arc from 1	shortest path now
2		30	30
3		10	10

Thus the program accepts the length of the direct connection between nodes 1 and 3 as the shortest path length that has been found from 1 to 3 so far. The cycle is repeated for node 4. This does not have a direct connection to the starting node. Consequently, slashes are displayed in columns 3 and 4 of the table.

trial values of shortest paths from 1			
node no.	shortest path yet	length of arc from 1	shortest path now
2		30	30
3		10	10
4		/	/

The table thus contains data for the direct connections (or absences of direct connection) between the starting node and all the other nodes.

The program now selects node 3 as the closest to node 1 in the entire graph. The reasons are as follows. The arc length from node 1 to node 3 is shorter than any other arc length from node 1. Because an arc length cannot be negative, any path between two nodes must be longer than an initial portion of that path. Thus any path to a third node that goes through node 3 must be longer than the path to node 3 itself. Any path to a third node that goes through node 2 must be longer than the path to node 2. We have already found that the direct connection to node 2 is longer than the direct connection to node 3. Any path to node 2 must therefore be longer than the path we have found to node 3. Any path to a third node that goes through node 2 must therefore be longer still. We can apply the corresponding argument to as many nodes as there are nodes in the graph. This reasoning is the key to the algorithm when the path lengths from the starting node are different. We show later why it also works when two or more path lengths are equal.

Continuing with the problem in hand, the prompt changes to "the shortest path just found leads to node 3—press the space bar to delete it from the trial value table." You respond, and the table contracts accordingly. The prompt changes to "press space bar to append node 3 to the final value table." The path length to node 3 now is displayed as the sole element (so far) of a second table below the first table of trial values.

trial values of shortest paths from 1			
node no.	shortest path yet	length of arc from 1	shortest path now
2		30	30
4		/	/

node no.	best path length (final)
3	10

Now the program proceeds to examine the distances from node 1 to the remaining nodes, along paths that go through node 3. The prompt changes to: "press space bar to transfer from the 'shortest now' to the 'shortest yet' column." You respond, and the trial value table is readied for another round of activity. This consists of forming the new 'best path lengths so far' and inspecting the arc lengths from node 3. The shortest path to node 2 that had been found previously was 30 units long. The path length via node 3 is $10+15$ units. Consequently, the prompt changes to "the path length to node 2 via node 3 is shorter— press space bar to update and continue."

The program deals with node 4 in a corresponding fashion. The path via node 2 thus is $10+20=30$ units long. Because the "shortest path yet" left from the preceding cycle is nonexistent, the path length 30 becomes the "shortest now" for node 4. The results at this point show that, of the two nodes 2 and 4, node 2 is closer to node 1. The program prompts you, accordingly, to press the space bar to delete the data for node 2 from the trial value table. The two tables now are as follows:

trial values of shortest paths from 1			
node no.	shortest path yet	length of arc from 3	shortest path now
4	/	20	30

node no.	best path length (final)
3	10
2	25

The program now investigates the path lengths to the residual node 4, via the node that was just finalized. The shortest path to node 4 found previously is 30 units long. The path length via node 2 is $25+25=50$ units long. Consequently, the previous value is retained and the data for node 4 is deleted from the trial value table and appended to the final value table.

trial values of shortest paths from 1			
node no.	shortest path yet	length of arc from 3	shortest path now

node no.	best path length (final)
3	10
2	25
4	30

The program recognizes that the trial list has been emptied, and it prompts you to type R to reexamine the current graph, A to examine another graph, or Q to quit. If you type R, the screen clears, the graph that you have been working with reappears, and you are prompted to type the number that identifies the node from which you want the shortest paths computed next. If you specify node 3 for the graph we have been discussing, the first cycle constructs a trial value table containing /, 15, and 20 for the "shortest path now" to nodes 1, 2, and 4, respectively.

The program picks the path to node 2 as the shortest and transfers node 2 to the final value table. The next cycle (1) finds that node 1 still cannot be reached and displays a message to that effect, (2) finds that the path to node 4 is not shortened by going via node 2, and (3) transfers the data for node 4 to the final value table. The next cycle just has node 1 left in the trial value table and finds that there is no edge to it from node 4. The program prompts: "Remaining node(s) cannot be reached from node 3—press the space bar to continue."

If you type R again, in response to the next prompt, and request the path lengths from node 2, the first cycle produces a trial value table that contains the edge length from node 2 to node 4 and shows that there are no edges from node 2 to nodes 1 and 3. The data about the path to node 4 is transferred to the final value table. The next cycle finds that there are no edges from node 4 to nodes 1 and 3. Accordingly, the message that the remaining nodes cannot be reached from node 2 appears. If you request the path lengths from node 4 for the same graph, the first cycle finds that none of the other nodes can be reached from it and displays the message to that effect.

Now type A for another graph in response to the prompt for further action and D for another automatic demonstration. A graph that consists of 6 nodes is displayed.

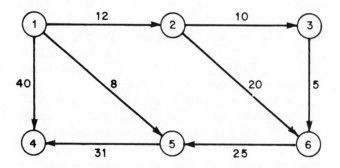

If you request the path lengths from node 1, the program produces the following table of shortest paths after five major cycles.

node no.	best path length (final)
5	8
2	12
3	22
6	27
4	39

If you type A and D again, for another automatic demonstration in response to the next two prompts, you see how a graph is treated when it contains nodes that are the same distance from the starting node. The following graph is displayed:

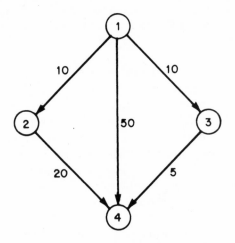

Suppose you request the shortest paths from node 1. The program shows that the arc lengths from the starting node to nodes 2, 3, and 4 are 10, 10, and 50, respectively. Two nodes tie for the shortest path at this juncture. The program arbitrarily picks the first of them and transfers node 2 to the final table. Then it examines the paths to nodes 3 and 4 via node 2. The path length to node 3 is unaffected. The path length to node 4 is reduced to $10 + 20 = 30$.

Now you can see from the graph that even though nodes 2 and 3 are the same distance from node 1, the path via node 2 is longer than the path via node 3. The fact that node 2 was picked before node 3 does not matter, however. Following the second pass through the trial table, node 3 is moved to the final value table. Then the program checks if paths through node 3 provide any improvement. For the only node left in the trial table, that is, node 4, the path length through node 3 is $10 + 5 = 15$. This is less than the value 30 left from the preceding cycle. The trial value table is updated and closed out, and the final value table is expanded to show that the best path lengths to nodes 2, 3, and 4 are 10, 10, and 15, respectively.

Three more preset demonstrations show the behavior (1) when paths of equal length go in opposite directions between two nodes, (2) when paths of unequal length join two nodes, (3) when there are 18 nodes, and (4) when the graph contains islands that cannot be reached from each other. Then typing D again merely restarts the cycle of automatic demonstrations after the sixth preset graph has been used. If you type I to input a new graph, the program prompts for the number of points, which must be in the range 3 to 20. To set up the graph with the structure shown on page 246, but with all the distances doubled, you begin by entering the value 4 for the number of nodes. The program prompts you to use the arrow keys to position node 1 and then to press the Enter key to label it. A pixel lights up on the screen to serve as a graphics cursor. By use of the arrow keys on the numeric key pad, you move this cursor to the point where you want node 1 displayed and press the Enter key. The numeral 1 appears at the point you selected, and a small circle is drawn around it. The 1 changes to 2 in the prompt. You move the cursor to the point where you want node 2 displayed and press the Enter key. The numeral 2 appears at the requisite position, surrounded by a circle. The 2 changes to 3 in the prompt. You set up node 3, and then node 4 similarly, by moving the cursor and pressing the Enter key as just described. Each time you press an arrow key, the graphics cursor moves one character position in the appropriate direction.

The program now prompts you to type A to define an arc, or C to start computing. You type A, and you are prompted to enter the node numbers of the start and end of the arc. You type 1,2 and press the Enter key. The program draws a line from the circle surrounding node 1 to the circle surrounding node 2. The line would go through the centers of the circles if it were continued. An arrowhead is drawn at the end of the line, pointing to node 2. The program prompts you to use the arrow keys to set the position where you want the edge length displayed and then to enter its value. You do that. The prompt to type A to define an arc or C to start computing reappears. You deal with all of the arcs in turn and then type C. After a brief pause, marked by the message "please wait—sorting data," the program prompts for the starting node, just as it did at the beginning of the automatic demonstration. You respond, and the algorithm is applied in the manner that has been described.

Program Variables

There are rather a lot of variables, but most are to take care of the display and step-by-step commentary.

HTP	The horizontal cursor position of the left edge of the tables
TW	The number of character positions that the tables occupy horizontally
RMW	The horizontal pixel position of the vertical line that separates the portions of the screen used to display the graph and the tables
BMW	The vertical pixel position of the horizontal line that separates the portions of the left half of the screen used to display the graph and the successive prompts
M$(i)	The ith line in the current prompt (i = 1 to 3)
R$	The one-character response to the prompts for certain options
NDS	The number of preset graphs
IDS	The ID number of the current preset graph (1 to 6)
NP	The number of nodes in the current graph
N	The index of successive nodes in several loops
X(n)	The horizontal pixel position of the nth node
Y(n)	The vertical pixel position of the nth node
D(n)	The shortest path length that, up to the current time in the execution of the algorithm, has been found between the starting node and node n
Q(n)	The pointer to the entry in the S, E, and L arrays (after the data in those arrays has been sorted in order of starting and ending nodes) for the first edge from node n (or from the first higher node with outward pointing edges if node n has none)
T(n)	The length of the edge between node n and the node from which edge lengths are being considered in the current cycle (this is set to a very large number if the nodes do not have a direct connection)
RT(n)	The nth node in the final value table
U(n)	The value of D(n) in the most recent cycle, retained for use in the display
HC	The horizontal position of the name of the node that is being displayed
VC	The vertical position of the name
KLIM	The maximum number of edges in a graph that the program accepts
KMAX	The number of edges in the graph under attention
K	The index of successive edges in the loops that process the edges
S(k),E(k)	The identification numbers of the nodes at the start and end of the kth edge of a graph

L(k)	The length (weight) of the kth edge
HCA	The horizontal position of the name of an edge that is being displayed
VCA	The corresponding vertical position
NS	The starting node from which shortest paths are required
FN DS$(n)	The string that displays the integer n without a leading space
F$	The string that monitors the progress of the algorithm. Initially it consists of NP zeros. The NSth character is set to 2 when the starting node NS has been specified. The kth character is changed to 1 when the shortest path has been found from the starting node to the kth node.
NINC	The number of nodes in the trial value table
J	The number of nodes in the final value table
NNEW	The node from which arc lengths are being considered
VCC	The vertical character position of the line of the trial table that is under consideration
ISOL	The switch that shows, at the end of a major cycle, if all remaining nodes are unreachable
LMIN	The interim value of the shortest path length for the nodes in the trial value table constructed most recently
FN C$(m)	The string of m cursor right codes
C1$,C3$	The cursor right code and a string of three of those codes
C4$,C8$	The strings of 4 and 8 cursor right codes
C14$,C22$	The strings of 14 and 22 cursor right codes
CM$	The string of HTP-1 cursor right codes to move from the left edge of the screen to the start of the tables
CD$	The cursor down code
DNEW	The path length from the starting node to the node that was transferred to the final value table most recently
TEMP	The path length via that node to the node under current attention
VCF	The vertical position of the current line of the final value table in the loop that displays this table
I	The text line number in the loops that clear the prompt area and display the prompts
R	The radius of the circle drawn around each node on the screen edges
PI	The constant π
PIBY2	The constant $\pi/2$
PIBY4	The constant $\pi/4$

PIBY6	The constant $\pi/6$
PI2	The constant 2π
ASP	The aspect ratio of the high-resolution pixels
S%	The style byte for the lines that form the edges of a graph
X,Y	The horizontal and vertical pixel positions of the graphics cursor
HC,VC	The corresponding horizontal and vertical positions of the text cursor
S$	The string that is typed after the graphics cursor has been positioned (this should be null for a node, and it should consist of 1 to 4 digits to specify the length of an edge)
V	The numerical value corresponding to S$
IDRN	The direction code for movement of the graphics cursor when a node or the length of an edge is being positioned on the screen
X1,Y1	The pixel coordinates of the node at one end of the edge of the graph that is being drawn
X2,Y2	The pixel coordinates of the node at the other end of the edge of the graph that is being drawn
TH	The angle that there would be between the edge and the horizontal if the pixels were square
THA	The angle that actually is seen because of the aspect ratio of the pixels
XS,YS	The pixel coordinates of one end of the edge where it reaches the circle around one node
XT,YT	The pixel coordinates of the other end of the edge

Program Notes

The program begins with some simple initialization, displays the main menu, and branches on the response. When an automatic demonstration is requested, the data that defines the structure of a graph and its appearance on the screen is read from some DATA statements and the graph is displayed. When the data that defines a graph is typed, some simple subroutines let the manipulation of the arrow keys determine the positions of the nodes and the displayed values of the edge lengths.

After a graph has been set up for an automatic demonstration or from the keyboard, the edge data in the S, E, and L arrays is sorted on E value (end node number) within S value (start node number). The directory Q is formed to allow the edges with the same S value to be found quickly in later operations. Thus, for the example of the tree on page 246, the successive entries in the S, E, and L arrays at this time are:

```
S   1   1   2   3   3
E   2   3   4   2   4
L  30  10  25  15  20
```

The entries in the Q array are 1, 3, 4, 6, and 6. The edges from node 1 occupy positions 1 through $3 - 1 = 2$ of the arrays. The edges from node 2 occupy positions 3 through $4 - 1 = 3$; that is, there is only one. The entries from node 3 occupy position 4 through $6 - 1 = 5$. The entries from node 4 occupy positions 6 through $6 - 1 = 5$, that is, there are none.

The string to control the cycle that builds the finalized list is initialized:

```
3200  F$=STRING$(NP,"0")
3210   MID$(F$,NS,1)="2"
```

NNEW is set to NS, the starting node. The arc lengths from the starting node to all the other nodes are transferred to the T array by a subroutine that first sets all the elements of the array T(n) to the value that signifies the absence of an edge. Then this value is overridden for the appropriate nodes by another loop.

```
8040  FOR K=Q(NNEW) TO Q(NNEW+1)-1
8050   T(E(K))=L(K)
8060  NEXT K
```

If NNEW has no outward pointing edges, then Q(NNEW) and Q(NNEW + 1) are equal. Consequently, the loop is not executed. Otherwise, the successive values of K in the range of the FOR statement point to successive elements of the arrays S, E, and L that contain data about outward-pointing edges from the node NNEW. The E(K) values are the actual node numbers of the ending nodes of the outward-pointing edges. Thus, for the first example considered in this section, the loop sets T(E(1)), that is, T(2), to 30. It sets T(E(2)), that is, T(3), to 10. T(4) is set to the very large number that shows the absence of an edge from NNEW.

By transferring the relevant data from the array T to the array D, the program uses the outward-pointing edges from the starting node as the first trial values of shortest paths to the nodes they reach. The loop goes through all the nodes in the graph and bypasses the starting node because the character that represents its status in F$ is 2.

```
4100 FOR N=1 TO NP
4110   IF MID$(F$,N,1)<>"0" THEN 4260
4120   D(N)=T(N)
. . .
4260 NEXT N
```

The other statements in the loop display the contents of the trial value table at this time.

The program then repeats a major loop that selects a node for the final value table and uses it to form a new set of values in the trial value table. The cycling ends when the string F$ shows that all the nodes have been transferred or the switch ISOL shows that all remaining nodes are unreachable.

```
5010   J=0
5020   ISOL=0
5030   WHILE INSTR(F$,"0") AND NOT ISOL
. . .
6670   WEND
```

The shortest path currently in the D(N) array is found by a simple loop that goes through all the nodes in the graph. The program refers to the F$ string to bypass the starting node and the nodes which have been finalized.

```
5040   LMIN=1E+20
5050   FOR N=1 TO NP
5060     IF MID$(F$,N,1)<>"0" OR D(N)>=LMIN THEN 5090
5070       NNEW=N
5080       LMIN=D(N)
5090   NEXT N
```

This loop leaves LMIN set to 10 to the power 20, that indicates the absence of a path, if none of the remaining nodes can be reached. In that circumstance, ISOL is turned on and the program jumps to the WEND statement at the end of the outer loop. Otherwise, NNEW identifies the node which ends the new shortest path. DNEW and the NNEWth character in the monitor string F$ are set appropriately. The count J of finalized nodes is increased, and the node number of the Jth entry in the final value table is set.

```
5120   DNEW=D(NNEW)
5130   MID$(F$,NNEW,1)="1"
. . .
5430   J=J+1
5440   RT(J)=NNEW
```

The lengths of the outward-pointing edges from the node NNEW to the unfinalized nodes are transferred to the array T(N). Then another loop works through the nodes and bypasses the nodes for which the monitor bit is not 0. For each node that still needs inspection, TEMP is set to (1) the path length from the starting node to node NNEW plus (2) the length of the edge from node NNEW to the node under consideration. The "best path length so far" to the node is reset if TEMP is less than the former value D(N).

```
6260   GOSUB 8000
. . .
6400   FOR N=1 TO NP
6410     IF MID$(F$,N,1)<>"0" THEN 6660
. . .
6450     TEMP=DNEW+T(N)
. . .
6470     IF TEMP<D(N) . . .
. . .
6530       D(N)=TEMP
. . .
6660   NEXT N
```

The detailed control structure and the further statements within the loop show the course of action by displaying appropriate prompts and entries in the trial value table. The switch ISOL is used to branch if all the remaining nodes cannot be reached.

Thus, the overall control structure of the main cycle can be summarized:

```
       [initialize F$, J, and ISOL]
5030   WHILE INSTR(F$,"0") AND NOT ISOL
       [select node NNEW to be transferred to final value table]
5130   MID$(F$,NNEW,1)="1"
       [remove data for node NNEW from trial value table]
       [redisplay the contracted trial value table]
       [append data for node NNEW to final value table]
       [redisplay the expanded final value table]
       [transfer from "new best" to "old best" columns on screen]
       [update "new best" values allowing for arcs from node NNEW]
6670   WEND
```

Suggested Activities

1. Simplify FSPATH by removing the sections that provide (a) the interactive display of a graph, (b) the step-by-step cycling, and (c) the display of the interim tabular data.

2. Expand the original or the simplified version of FSPATH to record the actual paths which are selected. (*Hint:* Include in the final value table a third column that points from each finalized node to its predecessor in the preferred path.)

3. Explore graphics techniques to emphasize and deemphasize the edges that pass into and out of attention.

```
1ØØ REM * FSPATH - find shortest path *
1Ø1 :
11Ø   ON ERROR GOTO 15ØØØ                                    'set error trap
12Ø   KEY OFF: SCREEN 2                                      'screen set up
13Ø   HTP=41                                                 'position of tables
14Ø   RMW=8*HTP-2Ø                                           'vertical divider
15Ø   TW=36                                                  'width of tables
16Ø   BMW=172                                                'top of prompt area
17Ø   NDS=6
18Ø   IDS=1                                                  'id of preset graph
19Ø   PI=3.14159                                             'pi
2ØØ   PIBY2=PI/2: PIBY4=PI/4                                 'multiples
21Ø   PIBY6=PI/6: PI2=2*PI
22Ø   ASP=5/12                                               'aspect ratio of pixels
23Ø   R=12                                                   'radius of node label
24Ø   S%=&H1111                                    'style byte for displayed graph
25Ø   NLIM=18
26Ø   KLIM=NLIM*(NLIM-1)
27Ø   DIM X(NLIM),Y(NLIM),D(NLIM),Q(NLIM+1),T(NLIM),RT(NLIM),U(NLIM)
28Ø   DIM S(KLIM),E(KLIM),L(KLIM)
29Ø   DEF FN DS$(N)=MID$(STR$(N),2)                          'removes spaces
3ØØ   DEF FN C$(M)=STRING$(M,28)                             'cursor right strings
31Ø   C1$=FNC$(1): C3$=FNC$(3): C4$=FNC$(4)
32Ø   C8$=FNC$(8): C14$=FNC$(14): C22$=FNC$(22)
33Ø   CM$=FNC$(HTP-2)                               'move past vertical median
34Ø   CD$=CHR$(31)                                           'cursor down code
399 :
4ØØ REM * initial screen set up *
41Ø   CLS                                                    'clear the screen
42Ø   LINE (RMW,Ø)-STEP(Ø,199)                               'rule off table area
43Ø   LINE (Ø,BMW)-STEP(RMW,Ø)                               'and prompt area
44Ø   M$(1)="Type D for automatic demonstration,"       'initial prompt message
45Ø   M$(2)="    I to input your own data,"
46Ø   M$(3)="      Q to quit:"
47Ø   GOSUB 95ØØ: R$=INPUT$(1)
48Ø   ON ((1+INSTR("DdIiQq",R$))\2) GOTO 5ØØ,1ØØØ,2ØØØØ       'branch
499 :
5ØØ REM * use preset data *
51Ø   READ NP                                                '# of nodes
52Ø   FOR N=1 TO NP
53Ø     READ HC,VC                                     'text coordinates of node
54Ø     GOSUB 1ØØØØ                                          'display it
55Ø   NEXT N
56Ø   READ KMAX                                              '# of arcs
57Ø   FOR K=1 TO KMAX
58Ø     READ S(K),E(K),L(K),HCA,VCA    'node #s of ends, weight, coords of label
59Ø     GOSUB 11ØØØ                                          'draw the arc
6ØØ     LOCATE VCA,HCA                                       'label it
61Ø     PRINT FN DS$(L(K))
62Ø   NEXT K
63Ø   DATA 4, 15,2, 3,8, 32,8, 15,16                                      :'1
64Ø   DATA 5, 1,2,3Ø,7,5, 1,3,1Ø,25,5, 3,2,15,19,7, 2,4,25,7,12, 3,4,2Ø,25,12
65Ø   DATA 6, 4,4, 19,4, 34,4, 4,1Ø, 19,1Ø, 34,1Ø                        :'2
66Ø   DATA 8, 1,2,12,1Ø,3, 1,4,4Ø,1,7, 1,5,8,13,7,   2,3,1Ø,27,3
67Ø   DATA   2,6,2Ø,28,7, 3,6,5,35,7, 5,4,31,12,11, 6,5,25,27,11
68Ø   DATA 4, 15,2, 3,8, 32,8, 15,16                                      :'3
69Ø   DATA 5, 1,2,1Ø,7,5, 1,3,1Ø,25,5, 1,4,5Ø,16,7, 2,4,2Ø,7,12, 3,4,5,25,12
7ØØ   DATA 3,13,2,3,12,23,12
71Ø   DATA 5,1,2,9Ø,7,1Ø,2,1,8Ø,8,4,1,3,2Ø,18,1Ø,2,3,12Ø,12,13,3,2,12Ø,12,13
72Ø   DATA 18, 3,2, 11,2, 19,2, 27,2, 35,2, 3,8, 11,8, 19,8, 27,8, 35,8
```

```
73Ø   DATA      3,14,11,14,19,14,27,14,35,14,3,2Ø,11,2Ø, 19,2Ø
74Ø   DATA 3Ø, 1,2,2Ø,6,1,2,3,3Ø,14,1,3,4,4Ø,22,1,4,5,5Ø,3Ø,1, 1,6,9ØØ,4,5
75Ø   DATA      6,1,9ØØ,4,5,2,7,8ØØ,12,5,3,8,7ØØ,2Ø,5,4,9,6ØØ,28,5,5,1Ø,5ØØ,36,5
76Ø   DATA      7,6,75,7,7,8,7,65,15,7,9,8,55,23,7,1Ø,9,45,31,7,6,11,4ØØ,4,11
77Ø   DATA      11,6,4ØØ,4,11,7,12,3ØØ,12,11,8,13,2ØØ,2Ø,11,9,14,1ØØ,28,11
78Ø   DATA      1Ø,15,8Ø,36,11,11,12,17Ø,6,13,12,13,16Ø,14,13,13,14,15Ø,22,13
79Ø   DATA      14,15,14Ø,3Ø,13,11,16,15,4,17,16,11,15,4,17,12,17,16,12,17
8ØØ   DATA      13,18,17,2Ø,17,17,16,14,7,19,18,17,13,15,19
81Ø   DATA 6, 4,4, 19,4, 34,4, 4,1Ø, 19,1Ø, 34,1Ø
82Ø   DATA 4, 1,4,4Ø,1,7, 1,5,8,13,7, 3,6,5,35,7, 5,4,31,12,11
83Ø   GOTO 2ØØØ
999 :
1ØØØ REM * user defined graph *
1Ø1Ø   M$(1)="Number of nodes (2-18): "
1Ø2Ø     GOSUB 95ØØ                                              'prompt
1Ø3Ø     INPUT; "",NP
1Ø4Ø       IF 1<NP AND NP<=NLIM THEN 12ØØ
1Ø5Ø         GOSUB 85ØØ                                          'invalid
1Ø6Ø         GOTO 1ØØØ
1199 :
12ØØ   PSET(16Ø,1ØØ)                                      'graphics cursor
121Ø   FOR N=1 TO NP                                      'loop through nodes
122Ø     M$(1)="Use arrow keys to position node"+STR$(N)          'prompt
123Ø     M$(2)="then press Enter key to label it"
124Ø     GOSUB 95ØØ
125Ø     GOSUB 12ØØØ                                    'position with arrow keys
126Ø     GOSUB 1ØØØØ                                     'display the node
127Ø   NEXT N
1399 :
14ØØ   K=Ø                                                      'arc #
141Ø   M$(1)="Type A to define an arc,"                         'prompt
142Ø   M$(2)="    C to start computing:"
143Ø     GOSUB 9ØØØ
144Ø     IF Z$="C" OR Z$="c" THEN 173Ø                    'jump to computation
145Ø     IF K<NP*(NP-1) THEN 15ØØ                               'check
146Ø     M$(1)="Limit reached on number of arcs -"        'too many arcs
147Ø     M$(2)="press C to compute, Q to quit:"
148Ø       GOSUB 9ØØØ                                            'prompt
149Ø       IF Z$="Q" OR Z$="q" THEN 2ØØØØ ELSE 173Ø             'branch
15ØØ   K=K+1                                                    'arc #
151Ø   M$(1)="Enter node #s of start & end: "
152Ø     GOSUB 95ØØ                                      'prompt for end node #s
153Ø     INPUT; "",S(K),E(K)
154Ø   IF Ø<S(K) AND S(K)<=NP AND Ø<E(K) AND E(K)<=NP
                         AND S(K)<> E(K)THEN 159Ø                'validate
155Ø     M$(1)="invalid node(s) - press space bar and"   'diagnostic
156Ø     M$(2)="try again"
157Ø     GOSUB 9ØØØ
158Ø     GOTO 141Ø                                              'retry
159Ø     GOSUB 11ØØØ                                      'draw the arc
16ØØ   M$(1)="Use the arrow keys to set position of"     'prompt to
161Ø   M$(2)="label, then type it and press Enter."       'position
162Ø   M$(3)="Use positive integers 1-9999"
163Ø     GOSUB 95ØØ                                              'the label
164Ø   GOSUB 12ØØØ                                    'use arrow keys to position it
165Ø   IF 1<V AND V<1ØØØØ AND V=V\1 THEN 171Ø                   'validate
166Ø   M$(1)="please use a positive integer less"        'diagnostic
167Ø   M$(2)="than 1ØØØØ - press space bar and"
168Ø   M$(3)="start again"
169Ø     GOSUB 9ØØØ
```

```
17ØØ    GOTO 72ØØ
171Ø    L(K)=V                                                        'weight
172Ø    GOTO 141Ø
173Ø    KMAX=K                                                        '# of arcs
1999 :
2ØØØ REM * sort the arc data and form the directory *
2Ø1Ø    M$(1)="Please wait - sorting data"
2Ø2Ø    GOSUB 95ØØ                                              'message area
2Ø3Ø    FOR K=KMAX-1 TO 1 STEP -1            'sort arcs by start & end node #s
2Ø4Ø    SW=-1
2Ø5Ø    FOR J=1 TO K
2Ø6Ø     IF S(J)<S(J+1) OR S(J)=S(J+1) AND E(J)<E(J+1) THEN 211Ø
2Ø7Ø      SW=Ø
2Ø8Ø      SWAP S(J),S(J+1)
2Ø9Ø      SWAP E(J),E(J+1)
21ØØ      SWAP L(J),L(J+1)
211Ø    NEXT J
212Ø    IF SW THEN 214Ø
213Ø    NEXT K
214Ø    N=1                                                  'form the directory
215Ø    S(Ø)=Ø                                    'start nodes of fictional arcs
216Ø    S(KMAX+1)=Ø
217Ø    IF KMAX=Ø THEN 224Ø
218Ø    FOR K=1 TO KMAX+1                                        'loop througn arcs
219Ø     IF S(K-1)=S(K) THEN 223Ø              'jump if start node # unchanged
22ØØ      Q(N)=K                               'pointer to 1st arc from new node
221Ø      N=N+1
222Ø      IF N<=S(K) THEN 22ØØ                 'allow for nodes without exits
223Ø    NEXT K
224Ø    IF Q(NP)=Ø THEN Q(NP+1)=Ø ELSE Q(NP+1)=KMAX+1          'allow for final node
2999 :
3ØØØ REM * prompt for starting node and prepare to find paths *
3Ø1Ø    M$(1)="Enter the # of the node from which"                'prompt
3Ø2Ø    M$(2)="you want shortest paths computed: "
3Ø3Ø    GOSUB 95ØØ
3Ø4Ø    INPUT; "",NS                                          'starting node #
3Ø5Ø     IF 1<=NS AND NS<=NP THEN 32ØØ                         'jump if valid
3Ø6Ø      GOSUB 85ØØ                                           'invalid response
3Ø7Ø      GOTO 3Ø1Ø                                            'try again
32ØØ    F$=STRING$(NP,"Ø")                                      'monitor string
321Ø    MID$(F$,NS,1)="2"                                       'starting node
322Ø    NINC=NP-1                                    'length of trial value table
33ØØ    M$(1)="press space bar to use the arcs from"            'prompt to use arcs
331Ø    M$(2)="the starting node"+STR$(NS)                      'from starting node
332Ø    GOSUB 9ØØØ
333Ø    LOCATE 1,HTP                                            'trial value table
334Ø    PRINT "trial values of shortest paths from" NS " "          'headings
335Ø    LOCATE 3,1
336Ø    PRINT CM$ " node   shortest   length of   shortest"
337Ø    PRINT CM$ "  no.    path yet   arc from" NS TAB(HTP+28) "path now"
34ØØ    LINE (8*HTP-12,8*CSRLIN-44)-STEP(8*TW+12,Ø)             'topmost horizontal
341Ø    LINE -STEP(Ø,8*NINC+44)                                 'right side
342Ø    LINE -STEP(-8*TW-12,Ø)                                  'base
343Ø    LINE -STEP(Ø,-8*NINC-44)                                'left side
344Ø    LINE (8*HTP-12,8*CSRLIN-28)-STEP(8*TW+12,Ø)             '2nd horizontal
345Ø    LINE (8*HTP-12,8*CSRLIN-4)-STEP(8*TW+12,Ø)              '3rd horizontal
346Ø    LINE (8*HTP+3Ø,8*CSRLIN-28)-STEP(Ø,8*NINC+28)           'inside verticals
347Ø    LINE (8*HTP+112,8*CSRLIN-28)-STEP(Ø,8*NINC+28)
348Ø    LINE (8*HTP+21Ø,8*CSRLIN-28)-STEP(Ø,8*NINC+28)
3999 :
```

```
4000 REM * use arc lengths from starting node to immediate neighbours *
4010   NNEW=NS
4020     GOSUB 8000                              'find weights of arcs from starting node
4030   VCC=6                                            'text cursor for trial table
4100   FOR N=1 TO NP                                      'loop through nodes
4110     IF MID$(F$,N,1)<>"0" THEN 4260                 'bypass for starting node
4120     D(N)=T(N)                                    'arc length from starting node
4130     M$(1)="press space bar to copy length of arc"                  'prompt
4140     M$(2)="from node"+STR$(NS)+" to node"+STR$(N)+" into the trial"
4150     M$(3)="value table"
4160     GOSUB 9000
4170     LOCATE VCC,HTP                        'display node # and arc length
4180     PRINT USING " ##&"; N, C4$+C14$;
4190     IF T(N)<1E+20 THEN PRINT USING "####";T(N); ELSE PRINT "   /";
4200     M$(1)="press space bar to use it as the best"
4210     M$(2)="length to node"+STR$(N)+" so far"               'prompt to use as
4220     GOSUB 9000                                        'interim path length
4230     LOCATE VCC,HTP+29
4240     IF D(N)<1E+20 THEN PRINT USING "######";D(N); ELSE PRINT "     /";
4250     VCC=VCC+1                                                'text cursor
4260   NEXT N
4999 :
5000 REM * main cycle *
5010   J=0                                          '# of finalized nodes
5020   ISOL=0                                     'isolated node switch off
5030   WHILE INSTR(F$,"0") AND NOT ISOL         'non-isolated nodes unfinalized
5040     LMIN=1E+20                             'prepare to find shortest arc
5050     FOR N=1 TO NP                                  'loop through nodes
5060       IF MID$(F$,N,1)<>"0" OR D(N)>=LMIN THEN 5090'jump unless arc is shorter
5070       NNEW=N                                      'new closest node
5080       LMIN=D(N)                                    '& shortest arc
5090     NEXT N
5100     ISOL=(LMIN>=1E+20)
5110     IF ISOL THEN 6680
5120     DNEW=D(NNEW)                                'shortest arc length
5130     MID$(F$,NNEW,1)="1"                         'marker in monitor string
5199 :
5200 REM * close up tentative list *
5210     M$(1)="the shortest path just found leads to"
5220     M$(2)="node"+STR$(NNEW)+" - press space bar to delete it"
5230     M$(3)="from the trial value table"
5240     GOSUB 9000                               'prompt to close trial list
5250     LOCATE 6,HTP
5260     FOR N=1 TO NP                                  'loop through nodes
5270     LOCATE CSRLIN,HTP
5280       IF MID$(F$,N,1)<>"0" THEN 5360                 'jump if finalized
5290       PRINT USING " ##&"; N, C4$;                    'redisplay if not
5300       IF U(N)=0 THEN PRINT C14$;
                    ELSE IF U(N)<1E+20 THEN PRINT USING "######";  U(N), C8$;
                                  ELSE PRINT "   /" C8$;
5310       IF T(N)<1E+20 THEN PRINT USING "####&"; T(N), C3$;
                    ELSE PRINT "  /" C3$;
5320       IF D(N)<1E+20 THEN PRINT USING "######"; D(N);
                      ELSE PRINT "     /";
5330       IF INSTR(MID$(F$,N+1),"0")=0 THEN 5340'jump if remaining nodes finalized
5340       PRINT CM$ C3$ SPC(2) C4$ SPC(6) C8$ SPC(4) C3$ SPC(6)
5350       Z$=INPUT$(1)
5360     NEXT N
5370     LOCATE CSRLIN,HTP-1                 'erase previous bottom line of data
5380     PRINT TAB(80)
```

```
5390    VCF=CSRLIN+1                                              'cursor position
5400    LINE (8*HTP-12,8*VCF-8)-STEP(8*TW+12,0),0      'erase previous base line
5410    LINE (8*HTP-12,8*VCF-16)-STEP(8*TW+12,0)            'draw new base line
5420    NINC=NINC-1                                          'occupancy of trial table
5430    J=J+1
5440    RT(J)=NNEW
5599    :
5600 REM * append to finalized list *
5610    M$(1)="press space bar to append node"+STR$(NNEW)+" to"
5620    M$(2)="the final value table"
5630     GOSUB 9000                          'prompt to append to finalized list
5640    LOCATE VCF,HTP-1
5650    VIEW (8*HTP-15,8*VCF)-(639,8*VCF+15),0                            'erase
5660    VIEW
5670    PRINT " node no.     best path length (final)" TAB(80) " ";       'heading
5680    LINE (8*HTP-12,8*VCF-10)-STEP(8*TW+12,0)               'top horizontal
5690    LINE -STEP(0,8*J+20)                                       'right side
5700    LINE -STEP(-8*TW-12,0)                                      'base line
5710    LINE -STEP(0,-8*J-20)                                      'left side
5720    LINE (8*HTP-12,8*VCF+4)-STEP(8*TW+12,0)                'center vertical
5730    LINE (8*HTP+76,8*VCF-10)-STEP(0,8*J+20)              'rule under headings
5740    LOCATE VCF+2,1
5750    FOR JJ=1 TO J                               'loop through finalized nodes
5760    N=RT(JJ)                                          'node # for jth row
5770    PRINT USING "& ##&#######"; CM$+C1$, N, C22$, D(N);      'redisplay
5780     IF JJ<J THEN LOCATE CSRLIN+1,1
5790    NEXT JJ
5800    IF INSTR(F$,"0")=0 THEN 6670                           'jump if complete
5999    :
6000 REM * transfer from "new best" to "old best" *
6010    VCC=6                                          'cursor for trial table
6020    FOR N=1 TO NP                                      'loop through nodes
6030    IF MID$(F$,N,1)<>"0" THEN 6130                      'jump if finalized
6040    M$(1)="press space bar to transfer from the"        'prompt to transfer
6050    M$(2)="'shortest now' to the 'shortest yet'"    'from 4th to 2nd column
6060    M$(3)="column"                               'column in tentative list
6070     GOSUB 9000
6080    LOCATE VCC,HTP+8                                              'transfer
6090    U(N)=D(N)
6100    IF U(N)<1E+20 THEN PRINT USING "######";U(N); ELSE PRINT "      /";
6110    PRINT C8$ SPC(4) C3$ "        ";                            'erase
6120    VCC=VCC+1                                          'cursor position
6130    NEXT N
6199    :
6200 REM * form new "best lengths" *
6210    M$(1)="press space bar to form the new 'best"       'prompt to form
6220    M$(2)="path lengths so far'"                       'new best lengths
6230     GOSUB 9000
6240    LOCATE 4,HTP+24                             'update node # in heading
6250     PRINT STR$(NNEW) SPC(-(NNEW<10)  )                    'of column 3
6260    GOSUB 8000                          'get arc lengths from this node
6270    VCC=6                                          'cursor line
6280    M$(1)="press space bar to inspect arc lengths"                'prompt
6290    M$(2)="from node"+STR$(NNEW)
6300     GOSUB 9000
6399    :
6400    FOR N=1 TO NP                                    'loop through nodes
6410    IF MID$(F$,N,1)<>"0" THEN 6660          'jump if node has been finalized
6420    LOCATE VCC,HTP+22
6430    IF T(N)>=1E+20 THEN PRINT "    /";: GOTO 6450      'nodes are not connected
```

```
6440    PRINT USING "####"; T(N);                           'display arc length
6450    TEMP=DNEW+T(N)                 .                     'path using it
6460    M$(1)="the path length to node"+STR$(N)+" via node"+STR$(NNEW)
6470    IF TEMP<D(N) THEN 6510 ELSE IF D(N)>=1E+20 THEN 6600 . 'jump if shorter
6480     M$(2)="is not shorter - press space bar to"
6490     M$(3)="continue"
6500      GOTO 6540                                          'jump if not
6510     M$(2)="is shorter - press space bar to update"
6520     M$(3)="and continue"                               'prompt to update
6530     D(N)=TEMP                                           'update
6540     GOSUB 9000                                          'wait for key press
6550    LOCATE VCC,HTP+29                     'display updated or retained value
6560     IF D(N)<1E+20 THEN PRINT USING "######";D(N); ELSE PRINT "    /";
6570     GOTO 6650                                           'jump
6599 :
6600     M$(1)="node"+STR$(N)+" still cannot be reached -"   'prompt
6610     M$(2)="press space bar to continue"
6620     GOSUB 9000
6630     LOCATE VCC,HTP+29                         'show node still isolated
6640      PRINT "    /";
6650    VCC=VCC+1                                            'cursor line
6660   NEXT N
6670  WEND
6680   IF NOT ISOL THEN 7000                            'prompt if impasse
6690    M$(1)="Remaining node(s) cannot be reached"
6700    M$(2)="from node"+STR$(NS)+" - press the space bar"
6710    M$(3)="to continue"
6720    GOSUB 9000
6999 :
7000 REM * all shortest paths found - prompt for further action *
7010    M$(1)="Type R to re-examine,"                       'prompt
7020    M$(2)="    A for another graph,"
7030    M$(3)="    Q to quit:"
7040    GOSUB 9500: R$=INPUT$(1)
7050    ON ((1+INSTR("RrAaQq",R$))\2) GOTO 7100,7200,20000     'branch
7100    VIEW (RMW+1,0)-(639,199)              're-examine present graph
7110    CLS                                   'erase table area
7120    VIEW
7130    ERASE D,RT,U
7140    DIM D(NLIM),RT(NLIM),U(NLIM)
7150    GOTO 3000
7200    ERASE X,Y,D,Q,T,RT,U,S,E,L
7210    DIM X(NLIM),Y(NLIM),D(NLIM),Q(NLIM+1),T(NLIM),RT(NLIM),U(NLIM)
7220    DIM S(KLIM),E(KLIM),L(KLIM)
7230    IF IDS<NDS THEN 7260
7240     RESTORE                        'restore preset data if final set used
7250     IDS=0
7260    IDS=IDS+1                        'subscript of preset data samples
7270    GOTO 400
7999 :
8000 REM * get lengths of arcs from selected node *
8010    FOR N=1 TO NP                                       'loop through nodes
8020     T(N)=1E+20
8030    NEXT N
8040    FOR K=Q(NNEW) TO Q(NNEW+1)-1       'relevant portion of arc length table
8050     T(E(K))=L(K)               'length of arc to appropriate end node
8060    NEXT K
8070    RETURN
8499 :
8500 REM * invalid response *
```

```
8510  M$(1)="Invalid reply - press any key to go on"        'invalid response
8520  GOSUB 9000
8530  RETURN
8999 :
9000 REM * prompt for key press *
9010  GOSUB 9500
9020  Z$=INPUT$(1)                                          'display message
                                                            'wait for key press
9030  RETURN
9499 :
9500 REM * display a prompt *
9510  FOR I=1 TO 3                                          'clear message area
9520   LOCATE 22+I,1
9530    PRINT TAB(HTP-2);
9540   NEXT I
9550  FOR I=1 TO 3                                  'loop through message lines
9560   IF M$(I)="" THEN 9600                             'jump if line is nul
9570   LOCATE 22+I,1
9580    PRINT M$(I);                                        'display the line
9590    M$(I)=""                                                   'nullify it
9600   NEXT I
9610  RETURN
9999 :
10000 REM * display a node *
10010  IF S$=CHR$(13) THEN 10040
10020   LOCATE VC,HC                                  'erase typed characters
10030   PRINT SPC(LEN(S$)-1)
10040  LOCATE VC,HC                                        'display node label
10050   PRINT FN DS$(N)
10060  X(N)=8*HC-5                                'draw circle around 1 digit label
10070  Y(N)=8*VC-5
10080  IF N>9 THEN 10110
10090  CIRCLE (X(N),Y(N)),R
10100  RETURN
10110  CIRCLE (X(N),Y(N)),R,,PIBY2,3*PIBY2                     'left semicircle
10120  CIRCLE (X(N)+8,Y(N)),R,,3*PIBY2,PIBY2                  'right semicircle
10130  LINE (X(N),Y(N)-R*ASP)-STEP(R,0)                      'and join around
10140  LINE (X(N),Y(N)+R*ASP)-STEP(R,0)                       '2 digit label
10150  RETURN
10999 :
11000 REM * draw an arc *
11010  X1=X(S(K))                                     'coords of starting node
11020  Y1=Y(S(K))
11030  X2=X(E(K))                                       'coords of ending node
11040  Y2=Y(E(K))
11050  IF X1=X2 THEN TH=PIBY2 ELSE TH=ATN((Y2-Y1)/(X2-X1)) 'allow vertical arcs
11060  IF X2<X1 OR X2=X1 AND Y2<Y1 THEN TH=PI+TH          '2nd and 3rd quadrants
11070  IF X2>X1 AND Y2<Y1 THEN TH=PI2+TH                        '4th quadrant
11080  XS=X1+(R+2)*COS(TH)                              'coords of start of edge
11090  YS=Y1+(R+2)*ASP*SIN(TH)
11100  XT=X2-(R+2)*COS(TH)                                   'and end of edge
11110  YT=Y2-(R+2)*ASP*SIN(TH)
11120  LINE (XS,YS)-(XT,YT),,,S%                             'draw the edge
11130  IF XT=XS THEN THA=PIBY2
             ELSE THA=ATN((YT-YS)/(ASP*(XT-XS)))  'angle adjusted for aspect
11140  IF XT<XS OR XT=XS AND YT<YS THEN THA=PI+THA      '2nd & 3rd quadrants
11150  IF XT>XS AND YT<YS THEN THA=PI2+THA                    '4th quadrant
11160  LINE (XT,YT)-STEP(-R*COS(THA+PIBY6),-R*ASP*SIN(THA+PIBY6))      'arrow
11170  LINE (XT,YT)-STEP(-R*COS(THA-PIBY6),-R*ASP*SIN(THA-PIBY6))       'head
11180  RETURN
11999 :
```

```
12000 REM * set the position of a node or arc label *
12010   X=POINT(0)                                      'coords of graphics cursor
12020   Y=POINT(1)
12030   R$=INKEY$                                         'wait for key press
12040   ON LEN(R$)+1 GOTO 12030,12400,12100    'nul, non-arrow or arrow key
12100     IDRN=INSTR("HKMP",RIGHT$(R$,1))             'arrow direction code
12110     IF IDRN=0 THEN 12030                         'ignore invalid arrow
12120   PSET(X,Y),1-POINT(X,Y)                   'erase the graphics cursor
12130   ON IDRN GOTO 12140,12160,12180,12200,   'branch on arrow direction
12140     IF Y>7 THEN Y=Y-8 ELSE Y=BMW: GOTO 12160                     'up
12150       GOTO 12220
12160     IF X>7 THEN X=X-8 ELSE X=RMW: GOTO 12140                   'left
12170       GOTO 12220
12180     IF X<RMW-8 THEN X=X+8 ELSE X=0: GOTO 12200                'right
12190       GOTO 12220
12200     IF Y<BMW-8 THEN Y=Y+8 ELSE Y=0: GOTO 12180                 'down
12210       GOTO 12220
12220   PSET(X,Y),1-POINT(X,Y)                        'display the pixel
12230     GOTO 12030
12400   HC=1+INT(X/8)                          'text cursor position of label
12410   VC=1+INT(Y/8)
12420   S$=R$                                     'initialize the label string
12430   GOTO 12480                                 'jump for node setting
12440   LOCATE VC,HC                          'display the label typed so far
12450     PRINT S$
12460     R$=INPUT$(1)                                     'get a character
12470     S$=S$+R$                               'append character to label
12480   IF R$<>CHR$(13) THEN 12440    'redisplay label unless Enter pressed
12490     V=VAL(S$)                                       'store the value
12500   RETURN
14999 :
15000 REM * error trap *
15010   M$(1)="error"+STR$(ERR)+" in line"+STR$(ERL)      'error diagnostic
15020   M$(2)="press space bar to go on"
15030   GOSUB 9000                                         'display it
15040   RESUME 20000
19999 :
20000 REM * quit *
20010   LOCATE 1,1                                      'for Ok message
20020   ON ERROR GOTO 0                     'reset error trap address
20030   END                                                      'quit
```

4. Expand FSPATH to find the node that minimizes the shortest path length to all other nodes.

5.6 Some Further Directions

The future of "knowledge processing" that deals with multiply connected data is geared in large part to mechanical operations on trees and graphs. The use of trees also is central to a considerable amount of work on sorting, searching, and related processes, and particular attention has been given to binary trees and to some other special kinds of trees for those purposes. This section comments briefly on (1) the depiction and description of trees in general, (2) some features of binary trees, (3) certain other special trees, (4) structure tables, and (5) connectivities and weights in graphs.

For a start, it is useful to have a linear depiction of trees that avoids diagrams or conventions that represent the connections by numbers. One simple method is to list the nodes vertically in their top-to-bottom, left-to-right order of occurrence in the tree and use indention to show the generation to which a node belongs. A related style dispenses with indention and precedes each node by the number that specifies the generation. The tree shown diagrammatically on page 194 is represented in these styles by:

a	1a
b	2b
e	3e
f	3f
c	2c
d	2d
g	3g
h	3h
i	3i
j	3j

The first style is used quite frequently in technical writing, in lists of subject headings, tables of contents, and indexes, and in summarized classification schemes. The second style parallels the definition of structures in PL/I. Nested parentheses are used in another style that is more compact but less convenient for readers to follow. It depicts the tree on page 194 as:

a(b(e,f),c,d(g,h,i,j))

This style can be defined recursively as follows. The root node is followed by the parenthesized list of its subtrees in left-to-right order. A comma separates each item in this list from its successor. A subtree that consists of a single node is represented by the name of the node. Every other subtree is represented by its root node followed by the parenthesized list of the subtrees of that node.

In our discussion of trees so far, we have taken for granted the idea that the relation between adjacent nodes (that is, nodes which are joined by an edge) allows a direction to be assigned to the edge. This is true, for example, in parent-child relationships. When the edges are directed, the tree is said to be directed, too. A tree can be undirected, for example, if the nodes represent physical objects and the edges represent pieces of string that join particular pairs. When an undirected tree is stored in the computer, the nodes must be arranged in some order, albeit arbitrary. Recognizing the equivalence of different representations of the same undirected tree is an important problem when composite objects are identified in artificial intelligence studies. In the field of organic chemistry, the skeleton of carbon atoms in an acyclic hydrocarbon constitutes an undirected tree. Elaborate nomenclature conventions have been established to number the nodes in a unique manner by reference to the branching pattern of the molecule.

In our earlier discussion we have also taken for granted an interest in a natural order of the children of an individual parent. This is not always the case. For example, the nodes may represent people, and interest may be restricted to parent-child relationships without reference to the relative ages of a group of siblings.

Binary trees are used for many purposes, in part because they can be represented by a simple extension of the linked list principle. To facilitate the traversal of a binary tree, the conventions are extended in a *threaded tree* R as follows. If node x is childless in a tree P, then its left child in R identifies the predecessor of x in the standard order of traversal of R. If node x has no younger sibling in P, then the right child of x in R identifies the successor of x in the standard traversal of R.

In practice, these ideas get tied quite rapidly to specific details of implementation. For example, in one approach that is based on the use of linked lists, each node is represented by a row of a three-column array. One item in the relevant row is the name of the node or some other data that pertains to the node but does not include information about the node's connection to other nodes. Quite often, this item points to the data in another area of storage. The second and third items are the subscripts of the rows that represent the left and right children, respectively, when they exist. A thread is stored as the negative of the subscript of the row that represents the predecessor or successor node. This kind of linked representation is useful when the tree is subject to frequent and rapid changes.

An ordered set of objects can be represented by a binary tree in a way that allows members to be found more rapidly than in an ordinary linked list. Each node represents one of the objects. The nodes in its left (right) subtree represent the objects that precede (follow) the object represented by the parent node. This applies at every level of the tree. Finding whether a given object is in the set consists of a succession of tests that result in a match that ends the search or a move to the left or right subtree of the node that was just tested. By balancing the tree to provide a high measure of symmetry, the process is made very efficient. Procedures to update such trees by inserting and deleting nodes, and rebalancing when necessary, have been studied extensively. One kind of binary tree that is important at times has the property that every node either has two children or has none. This is called a complete (or full) binary tree.

When dealing with directed trees, the term "outdegree" is used for the number of edges directed outward from a node. A tree is called m-ary if the maximum outdegree of its nodes is m. An m-ary tree can be ordered or unordered. It is complete if the outdegree of every node is either m or 0. Several kinds of trees have been studied for particular purposes, such as improving the performance of programs to maintain ordered sets of data that are stored as trees. Thus, in a "2-3 tree" every interior node has two or three children and the path lengths from the root to all the leaves are equal. A binary search tree in which the heights of two siblings must either be equal or differ by 1 is called an AVL tree (after its inventor's initials). A B tree of order m is an m-ary search tree in which (1) the root has more than one child (or is the only node), (2) every other interior node has between int (m/2) and m children, and (3) the path lengths from the root to all the leaves are equal. A trie (rhymes with pi) is an m-ary tree in which the concatenation of a particular property of the nodes in the path from the root to a leaf is of interest. For example, the property may be a letter, and the paths then spell out the words in a vocabulary.

Although the use of structure tables has not been investigated as extensively as these other methods, it has the advantage of maintaining a close and obvious relation to the natural structure of the data. This can be of major importance to users whose primary concern is the subject matter of new applications. Such users may find the structure table approach to be immediately intelligible and readily adapted to specialized features of the data and its applications.

The representation and manipulation of graphs is another area of ever-increasing interest. Graphs may be directed or nondirected and weighted or unweighted. Many methods of representation can be used. Thus a directed unweighted graph of n nodes can be represented by an $n \times n$ adjacency matrix, in which the (i,j)th element is 1 if an edge joins nodes i and j and is 0 otherwise. This representation has the interesting property that the (i,j)th element in the kth power of the matrix is the number of k-step paths between nodes i and j. Weighted graphs can be represented by adjacency matrices in which the elements are actual weights. A variety of storage techniques, including linked lists, economize the space occupied by sparse matrices such as frequently occur in the representation of graphs.

When dealing with unweighted graphs, one class of algorithms systematically finds all the paths between two nodes. This traversal problem provides obvious scope for backtracking methods that use stacks. It includes the problem of escaping from a maze. The general definition of a graph does not require the existence of paths between all the nodes. The set of edges can even be empty, which will leave every node in complete isolation. Algorithms that simply find whether any path exists between two specified nodes, without providing the actual details, are related to the traversal problem.

A tree that connects all the vertices of a graph is called a spanning tree. Its total weight is the sum of the weights of its constituent edges. A minimum spanning tree minimizes the total weight. The problem of finding the minimum spanning tree arises, for example, when communications and transportation networks are being planned.

Several algorithms exist for the shortest path problem considered in Section 5.5. Other important graph problems deal with PERT diagrams used in industrial planning, with network flow, and with geometrical structures of interest in computer graphics and other fields.

Perhaps the most exciting aspect of this work, however, is the scope for novel representations and algorithms for graphs with special properties that result from the applications of computers to new fields of endeavor.

Suggested Activities

1. Suppose that the nodes of an undirected tree P correspond to objects that could not be distinguished if the edges were removed. Design a scheme to represent the tree P by a unique directed and ordered tree Q. Thus, you might select the node with the highest degree in P to be the root node of Q, or you might make the longest path in P the path from the root node to the leftmost leaf of Q. Tie-breaking conventions are needed, and the scheme will be recursive. This problem paraphrases a standard issue in chemical nomenclature.

2. Write a program to convert an arbitrary representation of the undirected tree P into the standard form Q that conforms to the conventions you designed in Activity 1.

3. Design a scheme to determine if two trees S and T that represent undirected trees in arbitrary arrangements are equivalent. Explore the possible benefit of first testing for nonequivalence without converting both trees to the standard form. (For example, they cannot be equivalent if they contain different numbers of nodes or if the number of nodes of degree m in one tree is not equal to the number of nodes of degree m in the other tree.) Write a program to implement the scheme you designed.

4. Suppose that you are dealing with undirected trees in which the nodes are completely distinguishable without reference to the edges. Design a scheme to represent these trees unambiguously by directed trees. Write a program to implement the scheme. Design a scheme to decide whether arbitrary representations of two of the trees are equivalent. Write a program to implement the scheme.

5. Design a scheme to determine whether one undirected tree can be a subtree of another (a) when the nodes can be distinguished and (b) when they cannot be. This covers the standard chemical substructure searching problem.

6. Suppose you have a set of unbranched directed trees in which a "status number" is associated with each node. For example, if the trees are entries in a bibliography, the subject headings, subheadings, sub-subheadings, and actual citations could have status numbers 1 to 4, respectively. Write a program to combine these trees into a single tree in which (a) the root node is the title of the bibliography (status number 0), (b) its children are the subject headings arranged alphabetically, (c) the children of a heading consist of citations that are not covered by subheadings and then by any subheadings that the heading covers, (d) the children of a subheading consist of citations that are not covered by sub-subheadings and then by any sub-subheadings that the subheading covers, and (e) the children of a sub-subheading are the citations that the sub-subheading covers. Make the program completely general to any number of fragments with different status numbers. This example is a realistic model of bibliography production.

7. Write a program to convert the structure table representation of a tree into a linked representation of the corresponding threaded binary tree. Write a program to perform the reverse transformation.

8. Write a program to demonstrate the technique to find the number of k-step paths between nodes of a graph by using the adjacency matrix, which is mentioned on page 270.

9. Write a program to list all paths between two given nodes in an unweighted graph. Avoid infinite loops around any cyclic paths that are encountered.

10. Write a program to find a minimum spanning tree of a weighted graph.

Bibliography

Aho, A., J. Hopcroft, and J. Ullman: *Data Structures and Algorithms,* Addison-Wesley, Reading, Mass., 1983.

Amsbury, W.: *Data Structures from Arrays to Priority Queues,* Wadsworth, Belmont, Calif., 1985.

Augenstein, M., and A. Tenenbaum: *Data Structures and PL/I Programming,* Prentice-Hall, Englewood Cliffs, N.J., 1979.

Dale, N., and S. C. Lilly: *PASCAL Plus Data Structures, Algorithms and Advanced Programming,* Heath, Lexington, Mass., 1985.

Goodman, S. E., and S. T. Hedetniemi: *Introduction to the Design and Analysis of Algorithms,* McGraw-Hill, New York, 1977.

Horowitz, E., and S. Sahni: *Fundamentals of Computer Algorithms,* Computer Science Press, Rockville, Md., 1978.

—— and ——: *Fundamentals of Data Structures,* Computer Science Press, Rockville, Md., 1982.

Knuth, D. E.: *Sorting and Searching,* Addison-Wesley, Reading, Mass., 1973.

Maurer, H. A.: *Data Structures and Programming Techniques,* Prentice-Hall, Englewood Cliffs, N.J., 1977.

Pfaltz, J. L.: *Computer Data Structures,* McGraw-Hill, New York, 1977.

Reingold, E. M., and W. J. Hansen: *Data Structures,* Little, Brown, Boston, 1983.

Sedgewick, R.: *Algorithms,* Addison-Wesley, Reading, Mass., 1983.

Singh, B., and T. L. Naps: *Introduction to Data Structures,* West Publishing, St. Paul, Minn., 1985.

Standish, T. A.: *Data Structure Techniques,* Addison-Wesley, Reading, Mass., 1979.

Stubbs, D. F., and N. W. Webre: *Data Structures with Abstract Data Types and Pascal,* Brooks/Cole, Los Angeles, 1985.

Tenenbaum, A., and M. Augenstein: *Data Structures Using PASCAL,* Prentice-Hall, Englewood Cliffs, N.J., 1981.

Tremblay, J. P., and P. G. Sorenson: *An Introduction to Data Structures with Applications,* 2d ed., McGraw-Hill, New York, 1984.

Wirth, N.: *Algorithms & Data Structures,* Prentice-Hall, Englewood Cliffs, N.J., 1976.

Bibliography

Index

About the Authors

MICHAEL P. BARNETT, trained as a chemist, is professor of computer and information science at Brooklyn College of the City University of New York. Algorithms and data structures have been important elements of his work in computer applications during his years as professor at MIT and other universities and in industry. He is the author of over 70 technical papers and six books on computer programming and typesetting.

SIMON J. BARNETT brings to this text the perspectives of a younger generation that has grown up in a world of computers and is a recent graduate in computer science at Cornell University.

On the following pages, a number of key programs
are reproduced in Cauzin Softstrip™ Form
as a convenience for the reader.

The programs are:

QUIKSORT.BAS	(Chapter 1)
MULDIREC.BAS	(Chapter 2)
LINKLIST.BAS	(Chapter 3)
VALPOST.BAS	(Chapter 4)
FSPATH.BAS	(Chapter 5)

4

5

QWIKSORT

MULDIREC

1 2 3

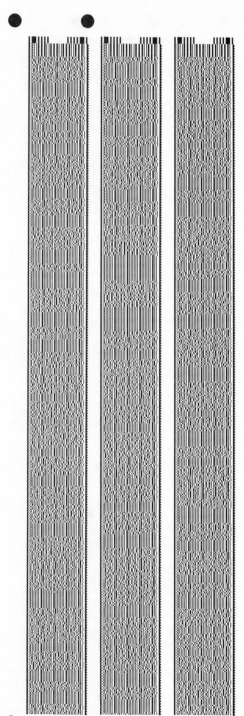

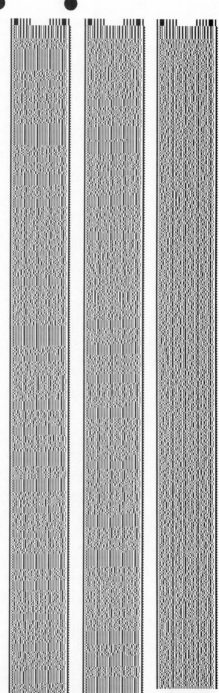

Softstrip

LINKLIST

1 2 3

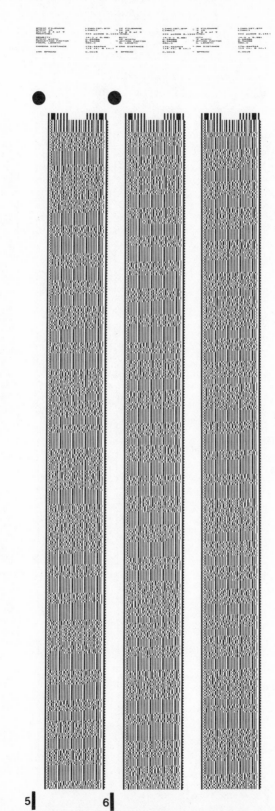

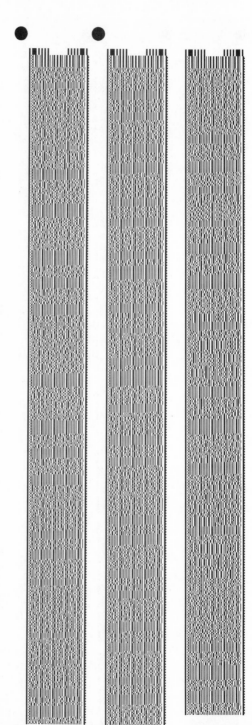

Softstrip
COMPUTER READABLE PRINT

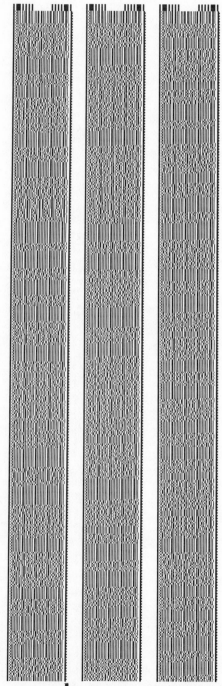

VALPOST

1 2 3

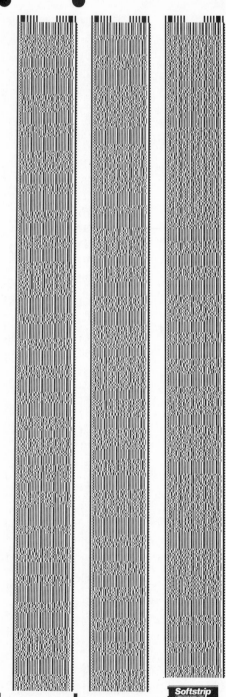

4

5

6

FSPATH

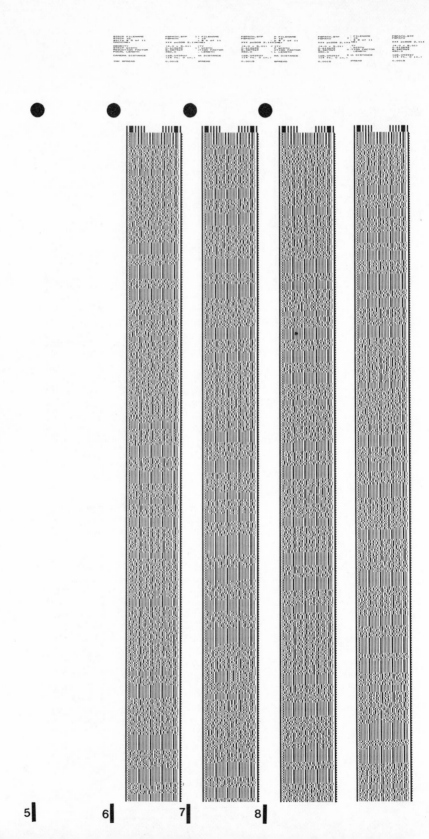

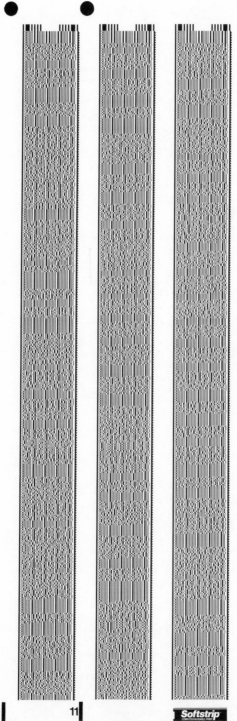